REUSE METHODOLOGY MANUAL

FOR SYSTEM -ON-A-CHIP DESIGNS

THIRD EDITION

Trademark Information

Synopsys and DesignWare are a registered trademarks of Synopsys, Inc.

Design Compiler is a trademark of Synopsys, Inc.

All other trademarks are the exclusive property of their respective holders and should be treated as such.

REUSE METHODOLOGY MANUAL

FOR SYSTEM -ON-A-CHIP DESIGNS

THIRD EDITION

By
Michael Keating
Synopsys, Inc.

Pierre Bricaud
Synopsys, Inc.

KLUWER ACADEMIC PUBLISHERS
Boston/Dordrecht/London

Distributors for North, Central and South America:
Kluwer Academic Publishers
101 Philip Drive
Assinippi Park
Norwell, Massachusetts 02061 USA
Telephone (781) 871-6600
Fax (781) 681-9045
E-Mail: kluwer@wkap.com

Distributors for all other countries:
Kluwer Academic Publishers Group
Post Office Box 322
3300 AH Dordrecht, THE NETHERLANDS
Telephone 31 786 576 000
Fax 31 786 576 474
E-Mail: services@wkap.nl

Electronic Services < http://www.wkap.nl>

Library of Congress Cataloging-in-Publication Data

Printed on acid-free paper.

Printed in the United States of America.

Table of Contents

Foreword

The electronics industry has entered the era of multimillion-gate chips, and there's no turning back. By the year 2001, Sematech predicts that state-of-the-art ICs will exceed 12 million gates and operate at speeds surpassing 600 MHz. An engineer designing 100 gates/day would require a hypothetical 500 years to complete such a design, at a cost of $75 million in today's dollars. This will never happen, of course, because the time is too long and the cost is too high. But 12-million gate ICs will happen, and soon.

How will we get there? Whatever variables the solution involves, one thing is clear: the ability to leverage valuable intellectual property (IP) through design reuse will be the invariable cornerstone of any effective attack on the productivity issue. Reusable IP is essential to achieving the engineering quality and the timely completion of multimillion-gate ICs. Without reuse, the electronics industry will simply not be able to keep pace with the challenge of delivering the "better, faster, cheaper" devices consumers expect.

Synopsys and Mentor Graphics have joined forces to help make IP reuse a reality. One of the goals of our Design Reuse Partnership is to develop, demonstrate, and document a reuse-based design methodology that works. The *Reuse Methodology Manual (RMM)* is the result of this effort. It combines the experience and resources of Synopsys and Mentor Graphics. Synopsys' expertise in design reuse tools and Mentor Graphics' expertise in IP creation and sourcing resulted in the creation of this manual that documents the industry's first systematic reuse methodology. The *RMM* describes the design methodology that our teams have found works best for designing reusable blocks and for integrating reusable blocks into large chip designs.

It is our hope that this manual for advanced IC designers becomes the basis for an industry-wide solution that accelerates the adoption of reuse and facilitates the rapid development of tomorrow's large, complex ICs.

Aart J. de Geus   *Walden C. Rhines*
Chairman & CEO   *President & CEO*
Synopsys, Inc.   *Mentor Graphics Corporation*

Preface to the Third Edition

The world of chip design has changed significantly since the second edition was published three years ago. In that time, silicon technology has gone through two generations, multi-million gate chips have gone from fringe to mainstream, and SoC has gone from the exotic to commonplace.

At the same time, the world of reuse has changed as well, prompting us to develop the third edition. From the perspective of 2002, many of the statements we made in 1999 now seem dated. Upon re-reading the second edition, it was obvious that the *RMM* needed to be updated with the many of the lessons learned in the last few years.

In one sense, though, the biggest change we have made in the *RMM* is also the biggest change we have seen in reuse in the last three years. Basically, we have changed the tense from future to present. Reuse is no longer a proposal; it is a solution practiced today by many, many chip designers. Likewise, the *RMM* is no longer aimed at promoting reuse, but describing changes in methodology for practising it. The *RMM* is now a chronicle of the best practices used by the best teams to develop reusable IP, and to use IP in SoC designs.

Alas, the change of tense was not as trivial a task as it sounds. In order to bring the *RMM* up to date, we have rewritten significant portions of the first eight chapters. Chapter 3 and Chapter 8 in particular have undergone significant revision. Chapter 5 has remained basically the same, with the addition of several important guidelines, and the modification of some existing guidelines to reflect current state-of-the-art.

Chapters 9 through 12 have had more modest updates to reflect current methodology. In particular, a full description of system level design and verification is beyond the scope of this book. Chapter 13 has been updated to include some comments from the design community on their perspective on reuse and SoC design.

In addition to the change in content, we have made one major editorial change. We have dramatically reduced the number of references to specific tools. Over the brief life of this book, we have found that tool names, tool vendors, and tool capabilities change so quickly that specific references quickly become out of date. Instead, we have focused on design and language issues, referencing the generic capabilities of current tools as appropriate.

We hope that readers will find the third edition a significant improvement over earlier editions.

May 1, 2002

Mike Keating   *Pierre Bricaud*
Mountain View, California   *Sophia Antipolis, France*

Acknowledgements

Over its brief life, the *RMM* has become very much a collaborative effort, with contributions from many different people from many different companies. The authors have had numerous conversations with engineers and managers, discussing their struggles with SoC design and reuse methodology, and the solutions they have developed to meet these challenges. These informal conversations have been extremely valuable in improving the content of the book. In particular, we would like to thank:

- Andre Kuntz, Jean-Claude Six, Neil Tebbutt and Christophe Dejean of Philips Semiconductor
- Pete Cummings and Christian Rousset of Texas Instruments
- Jacques-Olivier Piednoir and Arnold Ginetti of Cadence Design Systems

We also want to thank the following individuals and their companies who participated in developing the reuse-based SoC design examples in Chapter 13.

- Thierry Pfirsch of Alcatel
- Erich Palm of Atmel
- Albert Stritter and Yves Saboret of Infineon Technologies
- Tim Daniels of LSI Logic
- Louis Quere, Pierre-Guy Margueritte, Pierre Lieutaud, Patrick Rousseau, and Alain Rambert of Philips Semiconductors
- Thierry Bauchon and François Remond of STMicroelectronics

In addition, a number of key individuals played a very active role, reviewing the text, commenting, suggesting, and criticizing. In particular, we would like to express appreciation for the efforts of David Flynn, John Biggs, and Simon Bates of ARM.

A special thanks goes to Anwar Awad, Han Chen, Alan Gibbons, and Steve Peltan of Synopsys for their technical contributions, and to Jeff Peek for his great help in coordinating and editing this third edition.

We would like to thank the following people who made substantial contributions to the ideas and content of the first two editions of the *Reuse Methodology Manual*:

- Warren Savage, Ken Scott, Shiv Chonnad, Guy Hutchison, Chris Kopetzky, Keith Rieken, Mark Noll, and Ralph Morgan

- Glenn Dukes, John Coffin, Ashwini Mulgaonkar, Suzanne Hayek, Pierre Thomas, Alain Pirson, Fathy Yassa, John Swanson, Gil Herbeck, Saleem Haider, Martin Lampard, Larry Groves, Norm Kelley, Kevin Kranen, Angelina So, and Neel Desai

- Nick Ruddick, Sue Dyer, Jake Buurma, Bill Bell, Scott Eisenhart, Andy Betts, Bruce Mathewson

- David Flynn, Simon Bates, Ravi Tembhekar, Steve Peltan, Anwar Awad, Daniel Chapiro, Steve Carlson, John Perry, Dave Tokic, Francine Furgeson, Rhea Tolman and Bill Rogers.

Finally, we would like to thank Tim and Christina Campisi of Trayler-Parke Communications for the cover design.

To our wives,
Deborah Keating and Brigitte Bricaud,
for their patience and support

CHAPTER 1   *Introduction*

Silicon technology now allows us to build chips consisting of hundreds of millions of transistors. This technology has enabled new levels of system integration onto a single chip, and at the same time has completely revolutionized how chip design is done. The demand for more powerful products and the huge capacity of today's silicon technology have moved System-on-Chip (SoC) designs from leading edge to mainstream design practice. These chips have one, and often several, processors on chip, as well as large amounts of memory, bus-based architectures, peripherals, coprocessors, and I/O channels. These chips are true systems, far more similar to the boards designed ten years ago than to the chips of even a few years ago.

As chips have changed, so has the way they are designed. Designing chips by writing all the RTL from scratch, integrating RTL blocks into a top-level design, and doing flat synthesis followed by placement, no longer works for complex chips. Design reuse—the use of pre-designed and pre-verified cores—is now the cornerstone of SoC design, for it is the only methodology that allows huge chips to be designed at an acceptable cost, in terms of team size and schedule, and at an acceptable quality. The challenge for designers is not whether to adopt reuse, but how to employ it effectively.

This manual outlines a set of best practices for creating reusable designs for use in an SoC design methodology. These practices are based on our own experience in developing reusable designs, as well as the experience of design teams in many companies around the world. Silicon and tool technologies move so quickly that many of the details of design-for-reuse will undoubtedly continue to evolve over time. But the fundamental aspects of the methodology described in this book have become widely adopted and are likely to form the foundation of chip design for some time to come.

1.1   Goals of This Manual

Development methodology necessarily differs between system designers and processor designers, as well as between DSP developers and chipset developers. However, there is a common set of problems facing everyone who is designing complex chips:

- Time-to-market pressures demand rapid development.
- Quality of results, in performance, area, and power, are key to market success.
- Increasing chip complexity makes verification more difficult.
- Deep submicron issues make timing closure more difficult.
- The development team has different levels and areas of expertise, and is often scattered throughout the world.
- Design team members may have worked on similar designs in the past, but cannot reuse these designs because the design flow, tools, and guidelines have changed.
- SoC designs include embedded processor cores, and thus a significant software component, which leads to additional methodology, process, and organizational challenges.

In response to these problems, design teams have adopted a block-based design approach that emphasizes design reuse. Reusing macros (sometimes called "cores") that have already been designed and verified helps to address all of the problems above. However, in adopting reuse-based design, design teams have run into a significant problem. Reusing blocks that have not been explicitly designed for reuse has often provided little or no benefit to the team. The effort to integrate a pre-existing block into new designs can become prohibitively high, if the block does not provide the right views, the right documentation, and the right functionality.

From this experience, design teams have realized that reuse-based design requires an explicit methodology for developing reusable macros that are easy to integrate into SoC designs. This manual focuses on describing these techniques. In particular, it describes:

- How reusable macros fit into a SoC development methodology
- How to design reusable soft macros
- How to create reusable hard macros from soft macros
- How to integrate soft and hard macros into an SoC design
- How to verify timing and functionality in large SoC designs

In doing so, this manual addresses the concerns of two distinct audiences: the creators of reusable designs (macro designers) and chip designers who use these reusable blocks (macro integrators). For macro designers, the main sections of interest will be those on how to design soft macros and turn them into hard macros, and the other sections will be primarily for reference. For integrators, the sections on designing soft and hard macros are intended primarily as a description of what to look for in reusable designs.

SoC designs are made possible by deep submicron technology. This technology presents a whole set of design challenges including interconnect delays, clock and power distribution, and the placement and routing of millions of gates. These physical design problems can have a significant impact on the functional design of SoCs and on the design process itself. Interconnect issues, floorplanning, and timing design must be engaged early in the design process, at the same time as the development of the functional requirements. This manual addresses issues and problems related to providing logically robust designs that can be fabricated on deep submicron technologies and that, when fabricated, will meet the requirements for clock speed, power, and area.

SoC designs have a significant software component in addition to the hardware itself. However, this manual focuses primarily on the creation and reuse of reusable hardware macros. This focus on hardware reuse should not be interpreted as an attempt to minimize the importance in the software aspects of system design. Software plays an essential role in the design, integration, and test of SoC systems, as well as in the final product itself.

1.1.1 Assumptions

This manual assumes that the reader is familiar with standard high-level design methodology, including:

- HDL design and synthesis
- Design for test, including full-scan techniques
- Floorplanning, physical synthesis, and place and route

1.1.2 Definitions

In this manual, we will use the following terms interchangeably:

- Macro
- Core
- Block
- IP

All of these terms refer to a design unit that can reasonably be viewed as a stand-alone subcomponent of a complete SoC design. Examples include a PCI interface, a microprocessor, or an on-chip memory.

Other terms used throughout this manual include:

- **Subblock** – A subblock is a subcomponent of a macro (or core, block, or IP) that is too small or specific to be a stand-alone design component.
- **Soft macro** – A soft macro (or core, block, or IP) is one that is delivered to the integrator as synthesizable RTL code.
- **Hard macro** – A hard macro (or core, block, or IP) is one that is delivered to the integrator as a GDSII file. It is fully designed, placed, and routed by the supplier.

1.1.3   Virtual Socket Interface Alliance

The Virtual Socket Interface Alliance (VSIA) is an industry group working to facilitate the adoption of design reuse by setting standards for tool interfaces and design and documentation practices. VSIA has done an excellent job in raising industry awareness of the importance of reuse and of identifying key technical issues that must be addressed to support widespread and effective design reuse.

The working groups of the VSIA have developed a number of proposals for standards that are currently in review. To the extent that detailed proposals have been made, this manual attempts to be compliant with them.

Some exceptions to this position are:

- Virtual component: VSIA has adopted the name "virtual component" to specify reusable macros. We have used the shorter term "macro" in most cases.
- Firm macro: VSIA has defined an intermediate form between hard and soft macros, with a fairly wide range of scope. Firm macros can be delivered in RTL or netlist form, with or without detailed placement, but with some form of physical design information to supplement the RTL itself. We do not address firm macros specifically in this manual; we feel that it is more useful to focus on hard and soft macros.

1.2   Design for Reuse: The Challenge

An effective block-based design methodology requires an extensive library of reusable blocks, or macros. The developers of these macros must, in turn, employ a design methodology that consistently produces reusable macros.

This design reuse methodology is based on the following principles:

- The macro must be extremely easy to integrate into the overall chip design.
- The macro must be so robust that the integrator has to perform essentially no functional verification of internals of the macro.

1.2.1 Design for Use

Design for reuse presents specific challenges to the design team. But to be *reusable*, a design must first be *usable*: a robust and correct design. Many of the techniques for design reuse are just good design techniques:

- Good documentation
- Good code
- Thorough commenting
- Well-designed verification environments and suites
- Robust scripts

Engineers learn these techniques in school, but in the pressures of a real design project, they often take shortcuts. A shortcut may appear to shorten the design cycle, but in reality it just shifts effort from the design phase to the integration phase of the project. Initially, complying with good design practices might seem like an unnecessary burden. But these techniques speed the design, verification, and debug processes of any project by reducing iterations through the coding and verification loop.

1.2.2 Design for Reuse

In addition to the requirements above for a robust design, there are some additional requirements for a hardware macro to be fully reusable. The macro must be:

- **Designed to solve a general problem** – This often means the macro must be easily configurable to fit different applications.
- **Designed for use in multiple technologies** – For soft macros, this means that the synthesis scripts must produce satisfactory quality of results with a variety of libraries. For hard macros, this means having an effective porting strategy for mapping the macro onto new technologies.
- **Designed for simulation with a variety of simulators** – A macro or a verification testbench that works with only a single simulator is not portable. Some new simulators support both Verilog and VHDL. However, good design reuse practices dictate that both a Verilog and VHDL version of each model and verification testbench should be available, and they should work with all the major commercial simulators.
- **Designed with standards-based interfaces** – Unique or custom interfaces should be used only if no standards-based interface exists.
- **Verified independently of the chip in which it will be used** – Often, macros are designed and only partially tested before being integrated into a chip for verification, thus saving the effort of developing a full testbench for the design. Reusable designs must have full, stand-alone testbenches and verification suites that afford very high levels of test coverage.

- **Verified to a high level of confidence** – This usually means very rigorous verification as well as building a physical prototype that is tested in an actual system running real software.
- **Fully documented in terms of appropriate applications and restrictions** – In particular, valid configurations and parameter values must be documented. Any restrictions on configurations or parameter values must be clearly stated. Interfacing requirements and restrictions on how the macro can be used must be documented.

These requirements increase the time and effort needed for the development of a macro, but they provide the significant benefit of making that macro reusable.

1.3   The Emerging Business Model for Reuse

The economics of reuse have been changing almost as rapidly as the technology itself. A few years ago, there were many different questions:

- Should all designers design all their code to be reusable?
- How do you motivate design teams to make their designs available to other design teams in the company?
- What is the value of casual reuse—reuse of code that has not been explicitly designed for reuse?
- What blocks should be developed internally, and what blocks should be purchased from IP vendors?

To a large extent, these questions have been answered by the cumulative experience of the design community.

The first conclusion is that casual reuse provides about a 2-3x benefit. That is, reusing some code from a previous design, or using a block that has not been explicitly designed for reuse, costs about one-half to one-third the effort of designing it from scratch. This benefit is not nearly large enough to meet the needs of most SoC designs. Most blocks of most SoC designs must be explicitly designed for reuse.

A quick calculation shows why this is true. Good designers can design at a rate of about 100 gates per day, or about 30 lines of RTL code a day, if no reuse is employed. This figure has been roughly constant for about the last five years or so, and there is no indication that it will change much in the future. For a 100K gate ASIC (a typical 1990s design), this means 1000 designer-days, or a 5 person team for about a year. For a 10M gate design, this would mean 100,000 designer-days, or a 500 person team for one year. Even with a 2-3x improvement because of casual reuse, this is a prohibitive cost for most chips.

For this reason, companies have turned to explicit reuse as the solution to SoC design.

The second conclusion is that explicit design reuse requires a dedicated team. Designers who are working on a specific chip are not going to make their blocks reusable. Design for reuse takes significantly more time than design for a single use. Chip designers are always on a very aggressive schedule, and they simply are not going to jeopardize that schedule to design a block for future reuse. So companies have set up internal development groups dedicated to developing reusable blocks.

Thus, the vast majority of reusable blocks now come from dedicated internal teams or from third-party IP providers.

The third conclusion is that developing reusable blocks is very expensive. In the first edition of the *RMM*, we stated that designing a block for reuse costs 2-3x the cost of developing it for a single use. We now believe that is estimate was too low. To achieve the levels of functional verification, completeness of documentation, and support infrastructure required to make a macro genuinely reusable is significantly more than a 3x effort. For small blocks, 10x may be more realistic; for complex blocks such as processors the cost can be an order of magnitude higher still.

For this reason, chip design teams are buying whatever IP they can, and are developing reusable macros only when necessary, such as for those functions that are differentiating or unique to the end product. This buy-before-build has been complicated by the relative lack of high-quality, third-party IP. Most design teams who have used a lot of third-party IP have had at least some bad experiences, with functional bugs and poor documentation being major issues. But some third-party IP providers are consistently delivering very high quality IP that dramatically speeds time-to-market for SoC designs. The major processor providers, in particular, have had success in placing their cores in more and more designs, even in competition with internally developed processors.

The need to buy-before-build, and the relative scarcity of really good third-party IP has contributed to the trend of standardization in SoC designs. Increasingly, designers are selecting from a relatively small number of processors and buses as the foundation for their designs. They are also using standards-based interfaces, such as PCI-X, USB, and 1394, in order to assure interoperability of the final design. Thus, chips are converging rather than diverging in their basic structure, with differentiation being provided either by software or by specialized computational units.

This convergence of chip designs is reminiscent of the early days of the personal computer, when a large number of different architectures eventually reduced to two. This convergence of hardware platforms enabled a whole industry for software and for specialized plug-in hardware. It is likely that the convergence of chip architectures using reuse-based design will also enable new markets in software-differentiated products and very specialized, high-value hardware.

CHAPTER 2

The System-on-Chip Design Process

This chapter gives an overview of the System-on-Chip (SoC) design methodology. The topics include:

- Canonical SoC design
- System design flow
- The role of specifications throughout the life of a project
- Steps for system design process

2.1 A Canonical SoC Design

Consider the chip design in Figure 2-1 on page 10. We claim that, in some sense, this design represents a canonical or generic form of an SoC design. It consists of:

- A microprocessor and its memory subsystem
- On-chip buses (high-speed and low-speed) to provide the datapath between cores
- A memory controller for external memory
- A communications controller
- A video decoder
- A timer and interrupt controller
- A general purpose I/O (GPIO) interface
- A UART interface

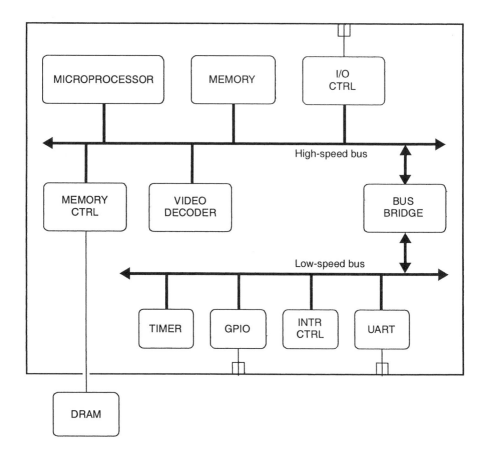

Figure 2-1   Canonical hardware view of SoC

This design is somewhat artificial, but it contains most of the structures and challenges found in real SoC designs. For example:

- The microprocessor could be anything from an 8-bit 8051 to a 64-bit RISC.

- The memory subsystem could be single- or multi-level, and could include SRAM
  and/or DRAM.

- The external memory could be DRAM (shown), SRAM, or Flash.

- The I/O controller could include PCI, PCI-X, Ethernet, USB, IEEE 1394, analog-
  to-digital, digital-to-analog, electro-mechanical, or electro-optical converters.

- The video decoder could be MPEG, ASF, or AVI.

- The GPIO could be used for powering LEDs or sampling data lines.

The design process required to specify such a system—to develop and verify the blocks, and assemble them into a fabricated chip—contains all the basic elements and challenges of an SoC design.

Real SoC designs are, of course, much more complex than this canonical example. A real design would typically include several sets of IP interfaces and data transformations. Many SoC designs today include multiple processors, and combinations of processors and DSPs. The memory structures of SoC designs are often very complex as well, with various levels of caching and shared memory, and specific data structures to support data transformation blocks, such as the video decoder. Thus, the canonical design is just a miniature version of an SoC design that allows us to discuss the challenges of developing these chips utilizing reusable macros.

2.2   System Design Flow

To meet the challenges of SoC, chip designers are changing their design flows in two major ways:

- From a waterfall model to a spiral model
- From a top-down methodology to a combination of top-down and bottom-up

2.2.1   Waterfall vs. Spiral

The traditional model for ASIC development, shown in Figure 2-2 on page 12, is often called a *waterfall model*. In a waterfall model, the project transitions from phase to phase in a step function, never returning to the activities of the previous phase. In this model, the design is often tossed "over the wall" from one team to the next without much interaction between the teams.

This process starts with the development of a specification for the ASIC. For complex ASICs with high algorithmic content, such as graphics chips, the algorithm may be developed by a graphics expert; this algorithm is then given to a design team to develop the RTL for the SoC.

After functional verification, either the design team or a separate team of synthesis experts synthesizes the ASIC into a gate-level netlist. Then timing verification is performed to verify that the ASIC meets timing. Once the design meets its timing goals, the netlist is given to the physical design team, which places and routes the design. Finally, a prototype chip is built and tested. This prototype is delivered to the software team for software debug.

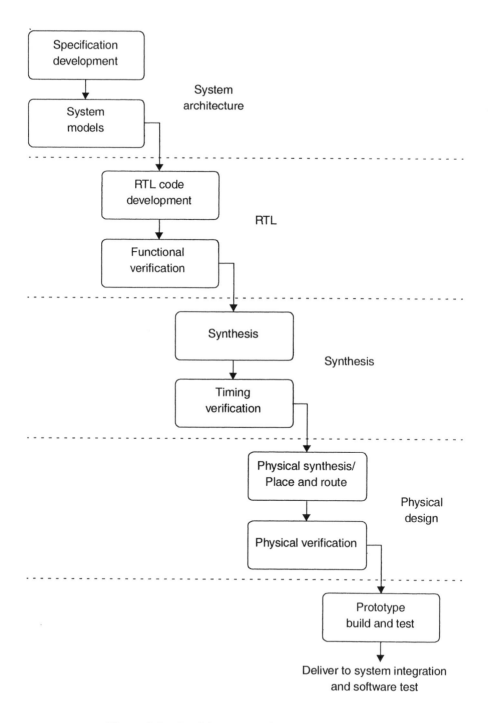

Figure 2-2   Traditional waterfall ASIC design flow

In most projects, software development is started shortly after the hardware design is started. But without a model of the hardware to use for debug, the software team can make little real progress until the prototype is delivered. Thus, hardware and software development are essentially serialized.

This flow has worked well in designs of up to 100K gates and down to 0.5 µm. It has consistently produced chips that worked right the first time, although often the systems that were populated with them did not. But this flow has always had problems because handoffs from one team to the next are rarely clean. For example, the RTL design team may have to go back to the system designer and tell him that the algorithm is not implementable, or the synthesis team may have to go back to the RTL team and inform them that the RTL must be modified to meet timing.

For large, deep submicron designs, this waterfall methodology simply does not work. Large systems have sufficient software content that hardware and software must be developed concurrently to ensure correct system functionality. Physical design issues must be considered early in the design process to ensure that the design can meet its performance goals.

As complexity increases, geometry shrinks, and time-to-market pressures continue to escalate, chip designers are turning to a modified flow to produce today's larger SoC designs. Many teams are moving from the old waterfall model to the newer *spiral development model*. In the spiral model, the design team works on multiple aspects of the design simultaneously, incrementally improving in each area as the design converges on completion.

Figure 2-3 on page 14 shows the spiral SoC design flow. This flow is characterized by:

- Parallel, concurrent development of hardware and software
- Parallel verification and synthesis of modules
- Floorplanning and place-and-route included in the synthesis process
- Modules developed only if a pre-designed hard or soft macro is not available
- Planned iteration throughout

In the most aggressive projects, engineers simultaneously develop top-level system specifications, algorithms for critical subblocks, system-level verification suites, and timing budgets for the final chip integrations. That means that they are addressing all aspects of hardware and software design concurrently: functionality, timing, physical design, and verification.

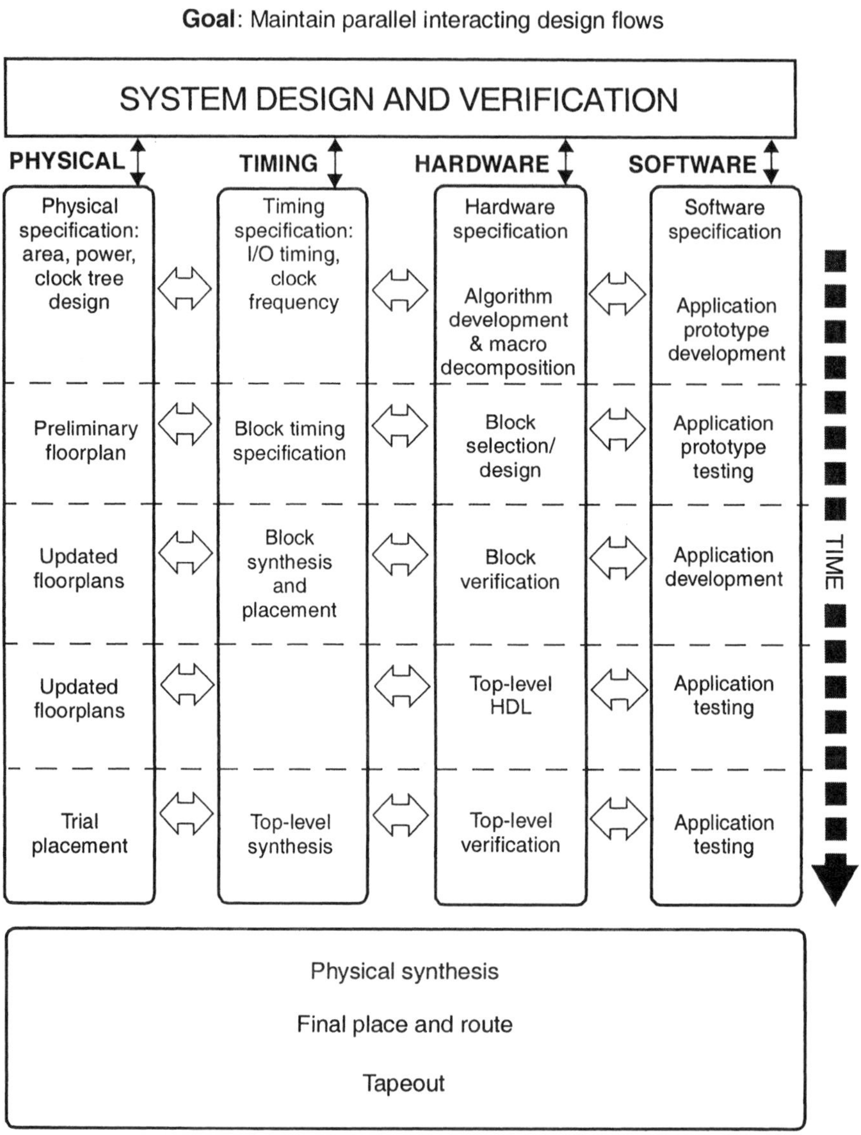

Goal: Maintain parallel interacting design flows

SYSTEM DESIGN AND VERIFICATION			
PHYSICAL	**TIMING**	**HARDWARE**	**SOFTWARE**
Physical specification: area, power, clock tree design	Timing specification: I/O timing, clock frequency	Hardware specification Algorithm development & macro decomposition	Software specification Application prototype development
Preliminary floorplan	Block timing specification	Block selection/ design	Application prototype testing
Updated floorplans	Block synthesis and placement	Block verification	Application development
Updated floorplans		Top-level HDL	Application testing
Trial placement	Top-level synthesis	Top-level verification	Application testing

TIME

| Physical synthesis |
| Final place and route |
| Tapeout |

Figure 2-3   Spiral SoC design flow

2.2.2   Top-Down vs. Bottom-Up

The classic top-down design process can be viewed as a recursive routine that begins with specification and decomposition, and ends with integration and verification:

1. Write complete specifications for the system or subsystem being designed.
2. Refine its architecture and algorithms, including software design and hardware/software cosimulation if necessary.
3. Decompose the architecture into well-defined macros.
4. Design or select macros; this is where the recursion occurs.
5. Integrate macros into the top level; verify functionality and timing.
6. Deliver the subsystem/system to the next higher level of integration; at the top level, this is tapeout.
7. Verify all aspects of the design (functionality, timing, etc.).

With increasing time-to-market pressures, design teams have been looking at ways to accelerate this process. Increasingly powerful tools, such as synthesis and emulation tools, have made significant contributions. Developing libraries of reusable macros also aids in accelerating the design process.

But, like the waterfall model of system development, the top-down design methodology is an idealization of what can really be achieved. A top-down methodology assumes that the lowest level blocks specified can, in fact, be designed and built. If it turns out that a block is not feasible to design, the whole specification process has to be repeated. For this reason, real world design teams usually use a mixture of top-down and bottom-up methodologies, building critical low-level blocks while they refine the system and block specifications. Libraries of reusable hard and soft macros clearly facilitate this process by providing a source of pre-verified blocks, proving that at least some parts of the design can be designed and fabricated in the target technology and perform to specification.

2.2.3   Construct by Correction

The Sun Microsystems engineers that developed the UltraSPARC processor have described their design process as "construct by correction." In this project, a single team took the design from architectural definition through place and route. In this case, the engineers had to learn how to use the place and route tools, whereas, in the past, they had always relied on a separate team for physical design. By going through the entire flow, the team was able to see for themselves the impact that their architectural decisions had on the area, power, and performance of the final design.

The UltraSPARC team made the first pass through the design cycle—from architecture to layout—as fast as possible, allowing for multiple iterations through the entire process. By designing an organization and a development plan that allowed a single group of engineers to take the design through multiple complete iterations, the team was able to see their mistakes, correct them, and refine the design several times before the chip was finally released to fabrication. The team called this process of iteration and refinement "construct by correction."

This process is the opposite of "correct by construction" where the intent is to get the design completely right during the first pass. The UltraSPARC engineers believed that it was not possible at the architectural phase of the design to foresee all the implication their decisions would have on the final physical design.

The UltraSPARC development project was one of the most successful in Sun Microsystems' history. The team attributes much of its success to the "construct by correction" development methodology.

2.2.4   Summary

Hardware and software teams have consistently found that iteration is an inevitable part of the design process. There is significant value in planning for iteration, and developing a methodology that minimizes the overall design time. This usually means minimizing the number of iterations, especially in major loops. Going back to the specification after an initial layout of a chip is expensive; we want to do it as few times as possible, and as early in the design cycle as possible.

We would prefer to iterate in tight, local loops, such as coding, verifying, and synthesizing small blocks. These loops can be very fast and productive. We can achieve this if we can plan and specify the blocks we need with confidence that they can be built to meet the needs of the overall design. A rich library of pre-designed blocks clearly helps here; parameterized blocks that allow us to make tradeoffs between function, area, and performance are particularly helpful.

In the following sections we describe design processes in flow diagrams because they are a convenient way of representing the process steps. Iterative loops are often not shown explicitly to simplify the diagrams. But we do not wish to imply a waterfall methodology—that one stage cannot be started until the previous one is finished. Often, it is necessary to investigate some implementation details before completing the specification. Rather, it is our intention that no stage can be considered complete until the previous stage is completed.

A word of caution: the inevitability of iteration should never be used as an excuse to short-change the specification process. Spending time in carefully specifying a design is the best way to minimize the number of iterative loops and to minimize the amount of time spent in each loop.

2.3   The Specification Problem

The first part of the design process consists of recursively developing, verifying, and refining a set of specifications until they are detailed enough to allow RTL coding to begin. The rapid development of clear, complete, and consistent specifications is a difficult problem. In a successful design methodology, it is the most crucial, challenging, and lengthy phase of the project. If you know what you want to build, implementation mistakes are quickly spotted and fixed. If you don't know, you may not spot major errors until late in the design cycle or until fabrication.

Similarly, the cost of documenting a specification during the early phases of a design is much less than the cost of documenting it after the design is completed. The extra discipline of formalizing interface definitions, for instance, can occasionally reveal inconsistencies or errors in the interfaces. On the other hand, documenting the design after it is completed adds no real value for the designer and either delays the project or is skipped altogether.

2.3.1   Specification Requirements

Specifications describe the behavior of a system; more specifically, they describe how to manipulate the interfaces of a system to produce the desired behavior. In this sense, specifications are largely descriptions of interfaces. Functional specifications describe the interfaces of a system or block as seen by the user of the systems. They describe the pins, buses, and registers, and how these can be used to manipulate data. Architectural specifications describe the interfaces between component parts and how transactions on these interfaces coordinate the functions of each block, and create the desired system-level behavior.

In an SoC design, specifications are required for both the hardware and software portions of the design. The specifications must completely describe all the interfaces between the design and its environment, including:

Hardware
- Functionality
- External interfaces to other hardware (pins, buses, and how to use them)
- Interface to SW (register definitions)
- Timing
- Performance
- Physical design issues such as area and power

Software

- Functionality
- Timing
- Performance
- Interface to HW
- SW structure, kernel

Traditionally, specifications have been written in a natural language, such as English, and have been plagued by ambiguities, incompleteness, and errors. Many companies, realizing the problems this causes, have started using executable specifications for some or all of the system.

2.3.2   Types of Specifications

There are two major techniques currently being used to help make hardware and software specifications more robust and useful: *formal specification* and *executable specification.*

- **Formal specifications** – In formal specifications, the desired characteristics of the design are defined independently of any implementation. Once a formal specification is generated for a design, formal methods such as property checking can be used to prove that a specific implementation meets the requirements of the specification. A number of formal specification languages have been developed, including one for VHDL called VSPEC [1]. These languages typically provide a mechanism for describing not only functional behavior, but timing, power, and area requirements as well. To date, formal specification has not been used widely for commercial designs, but continues to be an important research topic and is considered promising in the long term.

- **Executable specifications** – Executable specifications are currently more useful for describing functional behavior in most design situations. An executable specification is typically an abstract model for the hardware and/or software being specified. For high-level specifications, it is typically written in C, C++, or some variant of C++, such as SystemC or a Hardware Verification Language (HVL). At the lower levels, hardware is usually described in Verilog or VHDL. Developing these models early in the design process allows the design team to verify the basic functionality and interfaces of the hardware and software long before the detailed design begins.

  Most executable specifications address only the functional behavior of a system, so it may still be necessary to describe critical physical specifications—timing, clock frequency, area, and power requirements—in a written document. Efforts are under way to develop more robust forms of capturing timing and physical design requirements.

2.4   The System Design Process

Many chip designs are upgrades or modifications of an existing design. For these chips, the architecture can be reasonably obvious. But for SoC designs with significant new content, system design can be a daunting process. Determining the optimal architecture in terms of cost and performance involves a large number of complex decisions and tradeoffs, such as:

*   What goes in software and what goes in hardware
*   What processor(s) to use, and how many
*   What bus architecture is required to achieve the required system performance
*   What memory architecture to use to reach an appropriate balance between power, area, and speed

In most SoC designs, it is not possible, purely by analysis, to develop a system architecture that meets the design team's cost and performance objectives. Extensive modeling of several alternative architectures is often required to determine an appropriate architecture. The system design process consists of proposing a candidate system design and then developing a series of models to evaluate and refine the design.

The system design process shown in Figure 2-4 on page 22 employs both executable and written specifications to specify and refine a system architecture. This process involves the following steps:

1. **Create the system specification**

    The process begins by identifying the objectives of the design; that is, the *system requirements*: the required functions, performance, cost, and development time for the system. These are formulated into a *preliminary specification*, often written jointly by engineering and marketing.

2. **Develop a behavioral model**

    The next step is to develop an initial high-level design and create a *high-level behavioral model* for the overall system. This model can be used to test the basic algorithms of the system design and to show that they meet the requirements outlined in the specification. For instance, in a wireless communication design it may be necessary to demonstrate that the design can meet certain performance levels in a noisy environment. Or a video processing design may need to demonstrate that losses in compression/decompression are at an acceptable level.

    This high-level model provides an executable specification for the key functions of the system. It can then be used as the reference for future versions of the design. For instance, the high-level model and the detailed design of a video chip can be given the same input stream, and the output frames can be compared to verify that the detailed design products the expected result.

3. **Refine and test the behavioral model**

A verification environment for the high-level model is developed to *refine and test* the algorithm. This environment provides a mechanism for refining the high-level design, verifying the functionality and performance of the algorithm. If properly designed, it can also be used later to verify models for the hardware and software, such as an RTL model verified using hardware/software cosimulation. For systems with very high algorithmic content, considerable model development, testing, and refinement occurs before the hardware/software partitioning.

For instance, a graphics or multimedia system may be initially coded in C/C++ with all floating-point operations. This approach allows the system architect to code and debug the basic algorithm quickly. Once the algorithm is determined, a fixed-point version of the model is developed. This allows the architect to determine what accuracy is required in each operation to achieve performance goals while minimizing die area.

Finally, a cycle-accurate and bit-accurate model is developed, providing a very realistic model for implementation. In many system designs, this refinement of the model from floating-point to fixed-point to cycle-accurate is one of the key design challenges.

These multiple models are very useful when the team is using hardware/software cosimulation to debug their software. The behavioral model can provide simulation for most development and debugging. Later, the detailed, cycle-accurate model can be used for final software debug.

4. **Determine the hardware/software partition (decomposition)**

As the high-level model is refined, the system architects determine the *hardware/software partition*; that is, the division of system functionality between hardware and software. This is largely a manual process requiring judgment and experience on the part of the system architects and a good understanding of the cost/performance tradeoffs for various architectures. A rich library of pre-verified, characterized macros and a rich library of reusable software modules are essential for identifying the size and performance of various hardware and software functions.

The final step in hardware/software partitioning is to define the interfaces between hardware and software, and specify the communication protocols between them.

5. **Specify and develop a hardware architectural model**

Once the requirements for the hardware are defined, it is necessary to specify a detailed hardware architecture. This involves determining which hardware blocks will be used and how they will communicate. Memory architecture, bus structure and bus bandwidth can be critical issues. Most SoC chips have many different blocks communicating over one or more buses. The traffic over these buses and thus the required bandwidth can be very application dependent, so it becomes necessary to evaluate architectures by running substantial amounts of representative application code on them.

Running significant amounts of application code on an RTL-level design is often too time consuming to be practical. To address this problem, designers are using transaction-level models to model interfaces and bus behavior. By eliminating the detailed behavior of pins and signals on a bus or interface, these models can run considerably faster than RTL models, yet still give accurate estimates of performance. Many of the features of the SystemC language were developed explicitly to facilitate transaction-level modeling.

Determining the final hardware architecture consists of developing, testing, and modifying architectural-level models of the system until an architecture is demonstrated to meet the system requirements.

6. Refine and test the architectural model (cosimulation)

One of the classic problems in system design is that software development often starts only once the hardware has been built. This serialization of hardware and software development has led to many delayed or even cancelled projects.

The architectural model for the system can be used for hardware/software cosimulation. It provides sufficient accuracy that software can be developed and debugged on it, long in advance of getting actual hardware.

As the software content of systems continues to grow, hardware/software co-development and cosimulation will become increasingly critical to the success of SoC projects. Having fast, accurate models of the hardware will be key to this aspect of SoC design.

7. Specify implementation blocks

The output of the architectural exploration activity is a *hardware specification*: a detailed specification of the functionality, performance, and interfaces for the hardware system and its component blocks.

In its written form, the hardware specification includes a description of the basic functions, the timing, area, and power requirements, and the physical and software interfaces, with detailed descriptions of the I/O pins and the register map.

The architectural model itself functions as an executable specification for the hardware.

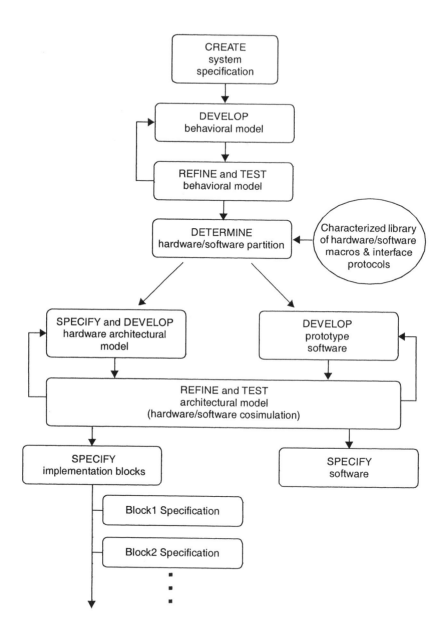

Figure 2-4   Top-level system design and recommended applications for each step

References

1. Ellsberger, Jan et al. *Sdl: Formal Object-Oriented Language for Communicating Systems.*
   Prentice Hall, 1997.

CHAPTER 3

System-Level Design
Issues: Rules and Tools

This chapter discusses system-level issues such as layout, clocking, floorplanning, on-chip busing, and strategies for synthesis, verification, and testing. These elements must be agreed upon *before* the components of the chip are selected or designed.

Topics in this chapter include:

- The standard model for design reuse
- Design for timing closure
- Design for verification
- System interconnect and on-chip buses
- Design for bring-up and debug
- Design for low power
- Design for test
- Prerequisites for reuse

3.1   The Standard Model

As more design teams use IP to do SoC designs, a consensus has emerged about some of the key aspects of reuse-based design. We call this view the "standard model" for design reuse.

In this standard model, the fundamental proposition is this: well-designed IP is the key to successful SoC design. No matter how good our SoC integration flow, if the blocks we are using are not designed well, the road to tapeout is long and very, very painful. On the other hand, well-designed IP can be integrated with virtually any (reasonably capable) SoC flow, and produce good results quickly.

In this chapter, we discuss the design guidelines for producing well-designed IP, as well as how to integrate well-designed IP into an SoC design. These guidelines are largely driven by the needs of the IP integrator and chip designer. In this sense they are basically system-level design guidelines.

In the next chapter, we discuss detailed coding guidelines, many of which are intended to implement the design guidelines discussed here.

There are some basic premises underlying all the guidelines in this book:

- **Discipline** – Building large systems (on a chip or otherwise) requires restricting the design domain to practices that consistently produce scalable, supportable, and easy to integrate designs.

- **Simplicity** – The simpler the design, the easier it is to analyze, to process with various tools, to verify, and to reach timing closure. All designs have problems; the simpler the design, the easier it is to find and fix them.

- **Locality** – Problems are easiest to find and solve when you know where to look. Making timing and verification problems local rather than global has a huge pay-off in reducing design time and improving the quality of a design. Careful block and interface design is essential for achieving this locality.

The authors, and many designers like us, learned these principles while designing large systems, and often learned them the hard way. For example (Mike speaking here), one of my first jobs was designing very large (hundreds of boards, each with hundreds of chips) ECL systems. When I arrived on the job, I was given a "green book" of how to do ECL system design. One of the rules was always to buffer inputs and outputs next to their edge connectors. This buffering essentially isolated the board, so that the backplane traces could be designed as transmission lines without knowing the details of how the daughter boards would load each signal. Essentially, it made both backplane and board design local (and relatively simple) design problems. The global problem, of designing a transmission line backplane with arbitrary stubs on the daughter boards, is totally intractable.

Similarly, on large chip designs, we can make block design into a local problem by carefully designing the interfaces. Good interfaces decouple internal timing and function (as much as possible) from the external behavior of the block, and thus from the timing and functional behavior of the other blocks. Thus, each block can be designed and verified in isolation. If the interfaces are consistent, then the blocks should plug and play; any remaining problems should be real, system-level design problems and not bugs in the block designs themselves.

This concept of locality is also fundamental to object-oriented programming. Here, classes are used to isolate internal data structures and functions from the outside world. Again, by carefully designing the interfaces between the class and the rest of the program, design and debug of key functionality can be isolated from the rest of the program. Software engineers have found that this approach to greatly facilitates design and debug of huge software systems, and greatly improves the quality of these systems.

There are two problems that dominate the SoC design process: achieving timing closure (that is, getting the physical design to meet timing), and functional verification. Design techniques for hard IP and soft IP are essentially the same regarding these two questions, relying on discipline, simplicity, and locality.

3.1.1   Soft IP vs. Hard IP

In the early days of reuse, there was some controversy over which blocks should be synthesized from RTL and which should be designed at the transistor level. In particular, some processor design teams insisted that processors must be full-custom designs in order to achieve their goals. As reuse-based SoC design has matured as a methodology, the roles of hard and soft IP have become quite clear.

Today, memory cells are designed by hand at the transistor level, and memory arrays are tiled from these cells using a compiler. Analog blocks, including digital-to-analog and analog-to-digital converters, as well as phase-locked loops are designed at least partially at the transistor level.

Virtually all other digital designs, however, start out as soft IP, and the RTL is considered the golden reference. Synthesis, placement, and routing then maps the RTL to gates, and the gates to GDSII. The process of mapping from RTL to GDSII is called *hardening*. The integration of the hardened block into the final chip is just the same as integrating any other hard IP, such as a memory.

In any SoC design, some blocks will be pre-hardened, and some will be synthesized as part of the final physical design of the chip, as shown in Figure 3-1 on page 26. But the basic design process for all digital blocks is basically the same.

The advantage of pre-hardened blocks is that they go through physical design once, and then can be used many times. For blocks such as processors that can be used without modification in many different chips, this can be a great advantage, since timing closure is completely predictable.

The advantage of synthesis-based design over full custom is that it provides a much faster implementation path for new designs. Full-custom digital design may be appropriate for the latest Intel processor, because huge teams of engineers can be applied to the problem. But for most embedded designs, economics dictates a different approach. The cost of porting a full-custom design to a new process is simply too expensive for any design that does not absolutely require it.

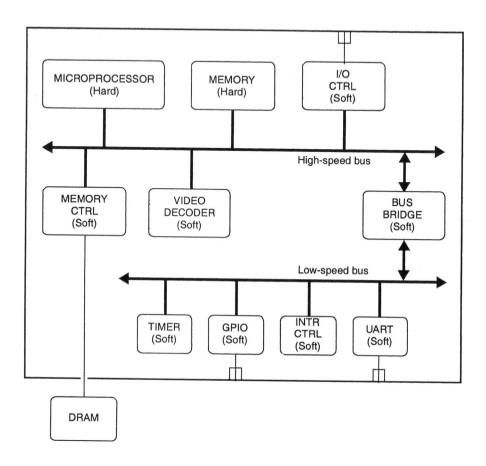

Figure 3-1   Sources of IP for the canonical design

3.1.2   The Role of Full-Custom Design in Reuse

In today's SoC design methodology, full-custom design plays a very specialized role by addressing those designs that do not lend themselves to synthesis, such as memories, analog blocks, and physical layer interfaces for high-speed communication protocols. For example, it is not possible (yet) to use synthesis to design a gigahertz Ethernet physical layer interface.

For everything else, synthesis-based standard cell design is the methodology of choice. This decision is driven by economics and the fact that any performance advantage of full-custom design has not turned into an advantage in the marketplace. But the question lingers of how much performance is being lost.

The performance penalty for semi-custom, standard cell based design appears to be quite small. In one recent design [1], a full-custom processor was redesigned using the design methodology described in this book. The results were:

- Maximum clock frequency was the same as for the full-custom design.
- Power was about 2x higher than the custom design, but no clock gating was used in the redesign. (Clock gating was used in the original design). Once a power synthesis tool was used to insert clock gating, the power was reduced to within a few percent of the original full-custom version. By using a power synthesis tool rather than hand-instantiated clock gating, both low power and full reusability were achieved.
- Area was initially about 7% larger than the full-custom design. However, all of this difference was found to be in a small, arithmetic operator. By replacing this operator with a full-custom version, the area for the entire processor was the same as for the full-custom version.

These results show why even the most aggressive processor designers are using full-custom techniques only for small portions of their designs. Processor designers tend to use synthesis for control logic and full custom only for data paths. The results above indicate that selective use of full custom only on small parts of the data path may produce the same results.

These observations lead to an interesting model for IP designers and integrators. We expect non-processor designs to avoid full-custom design completely. But for processor designs, integrators can use the RTL version of the processor as-is for rapid deployment in a new technology. For the most aggressive designs, they may selectively replace one or two key blocks with full-custom versions. This approach allows the integrator to balance time-to-market against performance, without incurring the full cost of a full-custom design.

The subject of the performance of synthesis-based design vs. full-custom design is explored in a series of papers [2,3].

3.2   Design for Timing Closure: Logic Design Issues

Timing and synthesis issues include synchronous or asynchronous design, clock and reset schemes, and selection of synthesis strategy.

3.2.1   Interfaces and Timing Closure

The proper design of block interfaces can make timing closure—both at the block level and system level—a local problem that can be (relatively) easily solved.

One of the major issues compounding the problem of timing closure for large chips is the uncertainty in interconnect wire delays. In deep submicron technologies, the wire delay due to wire load capacitance plus RC delay for the wires between the gates can be much larger than the intrinsic delay of the gate. Wire load models provide estimates of these wire delays for synthesis, but these are only estimates. As blocks become larger, the variance between the average delay (well estimated by the wire load model) and the actual delay on worst case wires can become quite large. To meet timing constraints, it may be necessary to increase the drive strengths of cells driving long wires. For very long wires, additional buffers must be inserted at intermediate points between the gates to assure acceptable rise and fall times as well as delays.

The problem is, of course, that the architect and designer do not know which wires will require additional buffering until physical design. If the designer has to wait until layout to learn that the design has to be modified to meet timing, then the project can easily suffer significant delays. If timing problems are severe enough to require architectural changes, such as increasing the pipeline depth, then other blocks, and even software, may be affected.

Timing-driven place and route tools can help deal with some of these timing problems by attempting to place critical timing paths so as to minimize total wire length. Physical synthesis, which combines synthesis with timing-driven placement, has taken great strides in managing the problems of achieving timing closure in deep submicron designs. But these tools cannot correct for fundamental architectural errors, such as an insufficient number of pipeline stages. And like most optimization tools, they work best on good designs—designs that are architected to make timing closure a straightforward, localized task.

Macro Interfaces

For macros, both inputs and outputs should be registered, as shown in Figure 3-2. This approach makes timing closure within each block completely local; internal timing has no effect on the timing of primary inputs and outputs of the block. Block A and Block B can be designed independently, and without consideration of their relative position on the chip. This design gives a full clock cycle to propagate outputs from one block to inputs of another. If necessary, buffers can to be inserted at the top level to drive long wires between blocks, without requiring redesign of Blocks A and B.

This kind of defensive timing design is useful in all large chip designs, but is essential for reuse-based SoC design. The IP designer does not know the timing context in which the block will be used. Output wires may be short or they may be many millimeters. Defensive timing design is the only way to assure that timing problems will not limit the use of the IP in multiple designs.

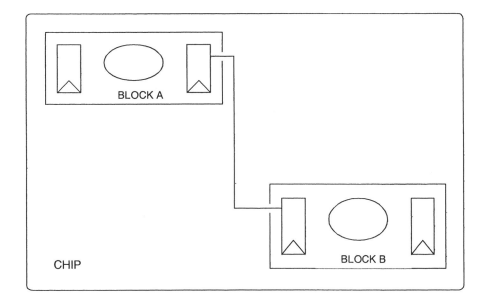

Figure 3-2   Registering inputs and outputs of major blocks

The major exception to this policy is the interface between a processor and cache memory. This interface is critical for high-performance designs, and usually requires special design. However, we prefer to think of the processor plus cache as being the true macro, and that the interface between this macro and the rest of the system should comply with the design guidelines mentioned above. For an example, see Figure 3-7 on page 44.

Subblock Interfaces

There is a corresponding design guideline for the subblocks of macros, as shown in Figure 3-3. For these designs, registering the outputs of the subblocks is sufficient to provide locality in timing closure. Because Block A is designed as a unit, and is relatively small, the designer has all the timing context information needed to develop reasonable timing budgets for the design.

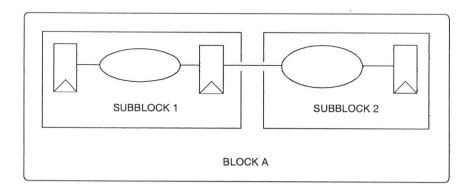

Figure 3-3   Registering outputs of subblocks

Because Subblock 1 is relatively close to Subblock 2, there is only a very small chance that the output wires from Subblock 1 to Subblock 2 will be long enough to cause timing problems [4]. Wire load estimates, synthesis results, and the timing constraints we give the physical design tools should all be accurate enough to achieve rapid timing closure in physical design.

There are several issues with this approach:

- When is a block large enough that we must register outputs?
- When is a block large enough that we must register both inputs and outputs?
- When can we break these rules, and how do we minimize timing risks when we do?

The first issue is reasonably straightforward: any block that is synthesized as a unit should have its outputs registered. Synthesis, and time budgeting for synthesis, is where we start striving for timing closure. This is where we want to start establishing locality in the timing of our designs.

The second issue has a similar answer: any block that is floorplanned as a unit should have its inputs and outputs registered. With blocks, especially reusable blocks, that are floorplanned as stand-alone units, we do not necessarily know how long the wires on its outputs and inputs will be. Registering all interfaces gives us the best chance of achieving timing closure for an arbitrary chip with an arbitrary floorplan. Consider our canonical design, Figure 3-1 on page 26. In some designs, we can assure that the memory and the video decoder will be close. But we would like to design the video decoder block so that it can be used in a wide variety of chip designs, including those where the memory is 10 or more millimeters away. For this reason, we want to register all the interfaces of the video decoder block.

We should violate these guidelines only when we absolutely need to, and only when we understand the timing and floorplanning implications of doing so. For instance, the PCI specification requires several levels of logic between the PCI bus and the first flop in the PCI interface block, for several critical control signals. In this case we cannot register all the inputs of the PCI bus directly; instead, we must floorplan the chip so that the PCI block is very close to the I/O pads for those critical control signals.

In summary, registering the interfaces to the major blocks of a design is the single most powerful technique for assuring timing closure. By localizing timing closure issues, the synthesis, timing analysis, and timing-driven place and route tools are allowed to work effectively.

Example: PCI vs. PCI-X

The original PCI specification required significant combinatorial logic between the input pin and the first register. This worked reasonably well at 33 MHz, but it made timing closure very difficult for 66 MHz. Having learned from this experience, the architects of the PCI-X specification designed it so that inputs were registered immendiately. As a result, timing closure on PCI-X at 66 MHz was dramatically easier that the PCI at the same speed, and closure was also achieved without difficulty at 133 MHz. These cases are shown in Figure 3-4 on page 32.

PCI (unregistered inputs)

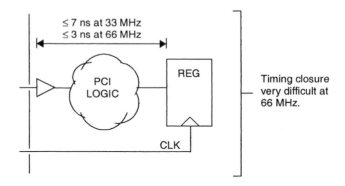

PCI-X (registered inputs)

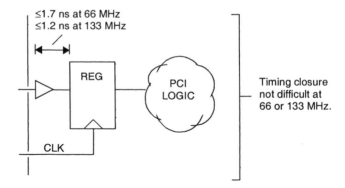

Figure 3-4   Unregistered vs. registered inputs for PCI and PCI-X

3.2.2   Synchronous vs. Asynchronous Design Style

Rule – The system should be synchronous and register (flip-flop) based. Latches should be used only to implement small memories or FIFOs. These memories and FIFOs should be synchronous and edge triggered. Exceptions to this rule should be made with great care and must be fully documented.

In the past, latch-based designs have been popular, especially for some processor designs. Multi-phase, non-overlapping clocks were used to clock the various pipeline stages. Latches were viewed as offering greater density and higher performance than register (flop) based designs. These benefits were sufficient to justify the added complexity of design.

Today, the tradeoffs are quite different. Deep submicron technology has made a huge number of gates available to the chip designer and, in most processor-based designs, the size of on-chip memory dwarfs the size of the processor pipeline. Also, with deep submicron design, delays are dominated by interconnect delay, so the difference in the effective delay between latches and flip-flops is minimal.

On the other hand, the cost of the increased complexity of latch-based design has risen significantly with the increase in design size and the need for design reuse.

Latch timing is inherently ambiguous, as illustrated in Figure 3-5 on page 34. The designer may intend data to be set up at the D input of the latch before the leading edge of the clock, in which case data is propagated to the output on the leading edge of clock. Or, the designer may intend data to be set up just before the trailing edge of the clock, in which case data is propagated to the output (effectively) on the trailing edge of the clock.

Designers may take advantage of this ambiguity to improve timing. "Time borrowing" is the practice of absorbing some delay by:

- Guaranteeing that the data is set up before the leading clock edge at one stage
- Allowing data to arrive as late as one setup time before the trailing clock edge at the next stage

The problem caused by the ambiguity of latch timing, and exacerbated by time borrowing, is that it is impossible by inspection of the circuit to determine whether the designer intended to borrow time or the circuit is just slow. Thus, timing analysis of each latch of the design is difficult. Over a large design, timing analysis becomes impossible. Only the original designer knows the full intent of the design. Thus, latch-based design is inherently not reusable.

For this reason, true latch-based designs are not appropriate for SoC designs. Some LSSD design styles are effectively register-based, however, and are acceptable if used correctly.

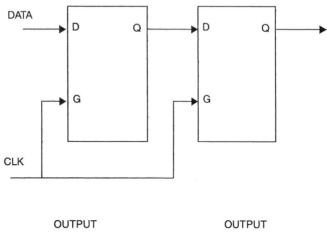

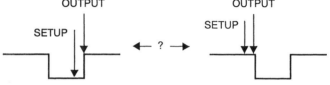

From which edge of the clock is data
propagated to the output?

Figure 3-5   Latch timing ambiguity

Example: Zycad Accelerators

Zycad Corporation developed a number of hardware accelerators during the 1980s
and 1990s. In one architecture, the design consisted of five chips operating together to
execute the simulation algorithm. These chips used a latch-based design and a four-
phase, non-overlapping clock. The pipeline was quite long, extending over all five
chips. To make timing, the team used a number of aggressive techniques. In some
stages, they knew the data would arrive at the flop before the leading edge of the
clock, and if the succeeding stage had a large amount of combinatorial logic, they
could allow data to settle as late as one setup time before the trailing edge of the
clock. Thus, although the system had a fixed clock, each stage of the pipeline had a
specification that could vary from:

```
clock_cycle + (2 * clock_width)
```

to:

```
clock_cycle - (2 * clock_width)
```

In addition, the team widened the second phase of the clock (to extend the trailing edge, and give some slow pipeline stages more setup time), and narrowed the third phase (pushing out the leading edge). Of course, the specification for this, and the definition of which specific stages required this extended timing, just resided in the heads of the designers and was never written down.

Five years later, it was time to update the design. Faster technologies and market pressure required moving the architecture to a smaller, faster process. Of course, by this time, all the engineers on the original design team had left the company. So a new team tried to guess the intent of their complex timing scheme. This rapidly proved to be an intractable problem. In fact, static timing analysis was never successfully performed on the new design. Instead, once silicon was received, clocks were adjusted until the system performed correctly. But it was never proved that the system worked at speed for worst-case silicon.

This is the kind of situation that keeps managers awake at night—one slow batch of silicon that could (possibly) shut down production of the company's sole product.

Fortunately, the Zycad accelerator shipped in small volumes, and no worst-case silicon was ever received from the chip vendor. But the experience with this machine shows the peril of aggressive latch-based designs, especially for designs intended for reuse.

3.2.3   Clocking

SoC designs almost always require multiple clock domains. For instance, in our canonical design (Figure 3-1 on page 26), the high-speed bus and the low-speed bus will have separate clocks. The interface/USB block may require yet another clock to match the data rate of its external interface. In addition, we may choose to shut down certain blocks at certain times to conserve power, and we may do this by shutting down the clock to those blocks. This effectively puts those blocks on a separate clock domain.

Clock Planning

Even one clock domain is a challenge; distributing a clock to tens or hundreds of thousands of registers with a skew low enough to avoid hold time problems will stress even the best clock tree synthesis tool. Every additional clock domain is an opportunity for disaster, since every time data crosses clock domains, there is an opportunity for metastability and corrupted data. For this reason, clock domains need to be managed very carefully.

Rule — The system clock generation and control logic should be separate from all functional blocks of the system. Keeping the clocking logic in a separate module allows the designer to modify it for a specific process or tool without having to touch the functional design.

Rule – The number of clock domains and clock frequencies must be documented. It is especially important to document:

- Required clock frequencies and associated phase-locked loops
- External timing requirements (setup/hold and output timing) needed to interface to the rest of the system
- Skew requirements between different, but related, clock domains (the difference between clock delays for Clock 1 and Clock 2.)

Guideline – Use the smallest possible number of clock domains. If two asynchronous clock domains interact, they should meet in a single module, which should be as small as possible. Ideally, this module should consist solely of the flops required to transfer the data from one clock domain to the other. The interface structure between the two clock domains should be designed to avoid metastability [5,6].

Guideline – If a Phase-Locked Loop (PLL) is used for on-chip clock generation, then some means of disabling or bypassing the PLL should be provided. This bypass makes chip testing and debug much easier, and facilitates using hardware modelers for system simulation.

Clock Delays for Hard Blocks

Clocks to hard macros present a special problem. Often they have a large insertion delay which must be taken into account during clock tree implementation.

Hard blocks should have their own clock net in a separate clock domain. This net can be used to compensate for the insertion delay, as described in Chapter 8. Once the insertion delay is determined, you can address it in either of two ways:

- *Eliminate* the delay using a PLL (to de-skew)
- *Balance* the clock insertion delay with the clock delay for the rest of the logic by giving the hard macro an early version of the clock.

3.2.4   Reset

Rule – The basic reset strategy for the chip must be documented. It is particularly important to address the following issues:

- Is it synchronous or asynchronous?
- Is there an internal or external power-on reset?
- Is there more than one reset (hard vs. soft reset)?
- Is each macro individually resettable for debug purposes?

There are advantages and disadvantages to both synchronous and asynchronous reset.

Synchronous reset:

- Advantage: Is easy to synthesize—reset is just another synchronous input to the design.
- Disadvantage: Requires a free-running clock, especially at power-up, for reset to occur.

Asynchronous reset:

- Advantage: Does not require a free-running clock.
- Advantage: Uses separate input on flop, so it does not affect flop data timing.
- Disadvantage: Is harder to implement—reset is a special signal, like clock. Usually, a tree of buffers is inserted at place and route.
- Disadvantage: Makes static timing analysis and cycle-based simulation more difficult, and can make the automatic insertion of test structures more difficult.

Note that regardless of whether synchronous or asynchronous reset is used, reset must be synchronously de-asserted in order to ensure that all flops exit the reset condition on the same clock. Otherwise, state machines can reset into invalid states.

Guideline – Asynchronous reset is preferred. Whatever the arguments about the relative advantages and disadvantages, asynchronous reset is by far the more commonly used method, especially in reusable designs. Ultimately, interoperability is more important that any small differences in ease of implementation.

3.2.5   Timing Exceptions and Multicycle Paths

In general, the standard model of reuse is for a fully synchronous system. Asynchronous signals and other timing exceptions should be avoided; they make chip-level integration significantly more difficult. The optimization tools—synthesis and physical synthesis—work best with fully synchronous designs. Once the clock frequency is defined, these tools can work to ensure that every path from flop to flop meets this timing constraint. Any exceptions to this model—asynchronous signals, multicycle paths, or test signals that do not need to meet this timing constraint—must be identified. Otherwise, the optimization tools will focus on optimizing these (false) long paths, and not properly optimize the real critical timing paths. Identifying these exceptions is a manual task, and prone to error. Our experience has shown that the fewer the exceptions, the better the results of synthesis and physical design.

3.3   Design for Timing Closure: Physical Design Issues

Once a design synthesizes and meets timing, timing closure becomes a physical design issue. Can we physically place and route the design so as to meet the timing constraints of the design? One of the keys to achieving rapid timing closure in physical design is to plan the physical design early.

3.3.1   Floorplanning

Rule – Floorplanning must begin early in the design process. The size of the chip is critical in determining whether the chip will meet its timing, performance, and cost goals. Some initial floorplan should be developed as part of the initial functional specification for the SoC design.

This initial floorplan can be critical in determining both the functional interfaces between macros and the clock distribution requirements for the chip. If macros that communicate with each other must be placed far apart, signal delays between the macros may exceed a clock cycle, forcing a lower-speed interface between the macros.

RTL Floorplanning

RTL floorplanning, which uses pre-synthesis RTL to generate a floorplan, is important in defining the physical partitioning of the design as well as managing the timing constraints (also known as budgeting). In many cases this partitioning results in a physical hierarchy that is different from the logical hierarchy. For example, a single physical partition may contain many logical blocks or macros. Also, for some timing-critical designs, parts of a single (logical) block may appear at different levels of the physical hierarchy. For instance, flops that are logically part of a larger block may need to be moved up in the physical hierarchy, in order to place them next to I/O pads.

The partitioning strategy includes the following:

• Define the logical hierarchies as soon as possible.
• Define the physical hierarchies based on physical relationships.
• Partition the constraints from a set of global constraints to several sets of local constraints.
• Partition scan chains and define I/O pins for them.

3.3.2   Synthesis Strategy and Timing Budgets

Rule – Overall design goals for timing, area, and power should be documented before macros are designed or selected. In particular, the overall chip synthesis methodology needs to be planned very early in the chip design process.

We recommend a bottom-up synthesis approach. Each macro should have its own synthesis script that ensures that the internal timing of the macro can be met in the target technology. This implies that the macro should be floorplanned as a single unit to ensure that the original wire load model still holds and is not subsumed into a larger floorplanning block.

Chip-level synthesis then consists solely of connecting the macros and resizing output drive buffers to meet actual wire load and fanout. To facilitate this, the macro should appear at the top level as two blocks: the internals of the macro (which are dont_touched) and the output buffers (which undergo incremental compile).

For large designs, floorplanning each block as a physical unit might not be appropriate for all blocks. Blocks might have to be part of the top level—especially the blocks that communicate with I/O cells. For a case study, see "Example: PCI vs. PCI-X" on page 31.

3.3.3   Hard Macros

Rule – A strategy for floorplanning, placing, and routing a combination of hard and synthesized soft macros must be developed *before* hard macros are selected or designed for the chip. Most SoC designs combine hard and soft macros, and hard macros are problematic because they can cause blockage in the placement and routing of the entire chip. Too many hard macros, or macros with the wrong aspect ratio, can make the chip difficult to place or route, or can create unacceptable delays on critical nets. Also, hard macros that use all routing layers will create a blockage problem, so some cores designed today leave a few layers for over-the-top routing. However, routing long nets can still be a problem, even for top-layer routing, if buffer locations are needed.

In Figure 3-6, a buffer with minimum wire delay is needed between Hard Macro 1 and Hard Macro 3. The appropriate place is already occupied by Hard Macro 2, so we would have to place the buffer to one side or the other of Hard Macro 2, which may not meet the timing requirements.

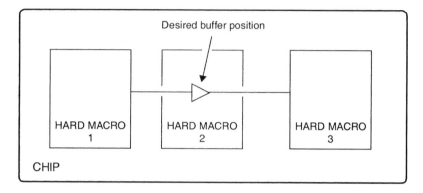

Figure 3-6   Buffer cannot be placed appropriately

3.3.4   Clock Distribution

Rule – The design team must decide on the basic clock distribution architecture for the chip early in the design process. The size of the chip, the target clock frequency, and the target library are all critical in determining the clock distribution architecture.

To date, most design teams have used a balanced clock tree to distribute a single clock (or set of clocks) throughout the chip, with the goal of distributing the clock with a low enough skew to prevent hold-time violations.

For large, high-speed chips, this approach can require extremely large, high-power clock buffers. These buffers can consume as much as half of the power in the chip and a significant percentage of the area.

3.4   Design for Verification: Verification Strategy

Design teams consistently list timing closure and verification as the major problems in chip design. For both of these problems, careful planning can help reduce the number of iterations through the design process. And for both problems, the principle of locality can help reduce both the number of iterations and the time each iteration takes, by making problems easier to find and to fix.

The objective of verification is to assure that the block or chip being verified is 100% functionally correct. In practice, this objective is rarely, if ever, achieved. In software, several defects per thousand lines of code is typical for new code [7,8]. RTL code is unlikely to be dramatically better.

We have found that the best strategy for minimizing defects is to do bottom-up verification; that is, to verify each macro or major block as thoroughly as possible before it is integrated into the chip. Finding and fixing bugs is easier in small designs. Then the major verification task at the chip level is to test the interaction between macros.

The major challenge in bottom-up verification is developing testbenches for the macro. For this reason, designers of large blocks intended for a single use often do cursory testing at the block level before integrating it into the chip. This approach may be more convenient, but it results in poorer verification, to the point where is simply does not work for large chips.

With modern high-level verification languages, creating testbenches at the macro level is considerably easier than before. For well-designed macros with clean, well-defined interfaces, these tools plus code coverage tools allow the designer to do very thorough verification at the macro level, as well as at the chip level.

Rule – The system-level verification strategy must be developed and documented before macro selection or design begins. Selecting or designing a macro that does not provide the modeling capability required for system-level verification can prevent otherwise successful SoC designs from completing in a timely manner. See Chapter 11 for a detailed discussion of verification strategies.

Rule – The macro-level verification strategy must be developed and documented before the design of macros and major blocks for the chip begins. This strategy should be based on bottom-up verification. Clear goals, testbench creation methodology, and completion metrics should all be defined. See Chapter 7 for a detailed discussion of macro-level verification.

The verification strategy also determines the kinds of testbenches required for system- or chip-level verification. These testbenches must accurately reflect the environment in which the final chip will work, or else we are back in the familiar position of "the chip works, but the system doesn't." Testbench design at this level is non-trivial and must be started early in the design process.

3.5   System Interconnect and On-Chip Buses

In the early days of reuse-based design, the wide variety of buses in common use presented a major problem. It seemed as if every chip design project had a unique bus, designed for optimum performance in that particular design. This approach made it difficult to use third-party IP or to reuse blocks from other projects within the company.

The solution to this problem is clearly to have a standard bus, allowing reusable blocks to be developed with a single interface that will allow it to be used in a wide variety of chips. Initially, committees across the industry and within major corporations attempted to develop standards to address this problem. These attempts met with modest success at best.

Meanwhile, design teams were under pressure to develop complex chips on very aggressive schedules. Design teams that were struggling with specialized buses, and with internally-developed processors, noticed that some of their colleagues were being much more successful in developing processor and bus-based designs. Teams that were adopting certain third-party processors and buses were finding that they had a rich set of peripherals, not to mention the processor itself, that were easy to connect and to get functioning correctly. The struggling teams quickly learned from this experience and started adopting the same approach. From this process a number of de facto standards have emerged. By far the most successful of these de facto standards are from ARM, MIPS, and IBM [9,10,11].

In a number of market segments, especially wireless, ARM processors have become the dominant standard. Similarly, in other product areas, MIPS processors have become standard, and in still others IBM's PowerPC processors are dominant. There are many other processors that have important niches in the industry, but these three processors families are clearly the most important, and are the ones that most affect how SoC designs are done.

As on-chip processors have converged, so has on-chip interconnect. The ARM processors are designed to work with ARM's AMBA bus, and there are many peripherals available both from ARM and third-party IP providers for this bus architecture. MIPS has its own EC bus; in addition, may SoC designs teams are using MIPS processors with the AMBA bus, so that they can take advantage of the large pool of available IP for AMBA. Finally, IBM has developed its CoreConnect bus and a large set of IP that works with it. The CoreConnect bus is only available to IBM's customers, however.

As a result, the AMBA bus has become the closest thing to an industry-wide standard for on-chip interconnect. CoreConnect is clearly the other key player in this area. Between them, they are driving the direction for on-chip buses.

Example: Integrating a Third-Party AGP Core

A few years ago, a startup was designing a graphics accelerator chip. They decided to use an AGP interface to memory, a standard approach for this kind of chip. Wanting to save design time, they purchased a synthesizable AGP core from a third-party IP vendor. Unfortunately, the AGP core's application interface (interface to the rest of the chip) did not at all match the chip's internal bus. As a result, the team had to design an interface block to adapt the AGP application interface to the system bus. This block ended up being nearly as complex as the AGP core itself, and was a major design effort. By the time the team has designed and debugged it, it was not clear that they had gained any advantage by purchasing the core, rather than designing one themselves that would integrate directly into the chip.

3.5.1   Basic Interface Issues

The version of our canonical design shown in Figure 3-7 on page 44 shows a common configuration for buses on an SoC design, using the AMBA bus as an example. A hierarchy of buses is used to reduce power while still meeting the bandwidth requirements of the various blocks in the system.

A private bus connects the processor to its cache and perhaps to other memories that it accesses frequently. For performance reasons, these memories are often designed for a specific process, rather than for general reuse. Thus, the processor, the private bus, and the memories form the reusable unit.

A high-speed bus (Advanced High-speed Bus or AHB in the case of AMBA) provides a high-performance interface between the processor and the other high-bandwidth blocks in the design. This bus is typically pipelined and supports several different kinds of transactions in order to maximize bandwidth. The AHB, for instance, has a two-stage pipeline, and supports split transactions as well as a variety of burst mechanisms.

A low-speed bus (Advanced Peripheral Bus or APB in the case of AMBA) provides a separate bus for low-speed peripherals and peripherals rarely accessed by the processor. By reducing the clock speed on the APB, it is possible to reduce the power consumption of the bus and the peripherals on it significantly.

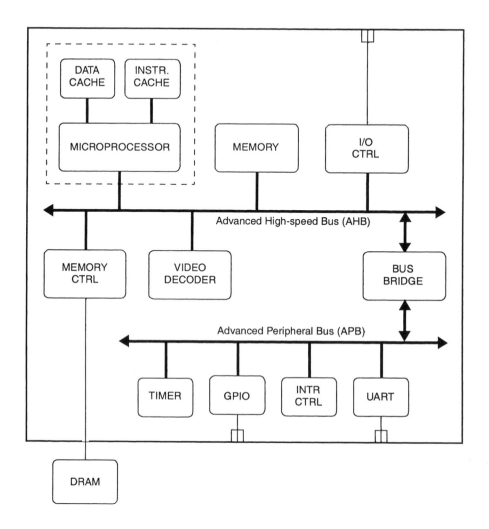

Figure 3-7   Canonical design with ARM on-chip buses

IBM's CoreConnect bus, shown in Figure 3-8, is similar to AMBA in concept, but supports a somewhat more complex protocol with a Processor Local Bus (PLB), On-Chip Peripheral Bus (OPB), and a Device Control Register Bus (DCR).

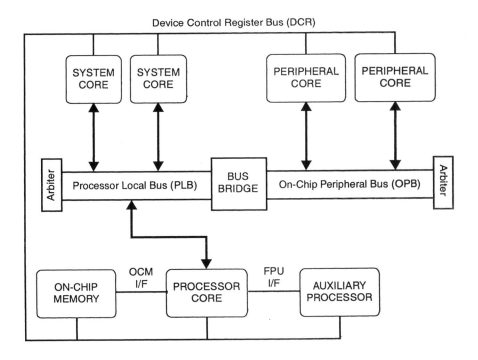

Figure 3-8   CoreConnect block diagram[1]

In very high performance designs, it may be necessary to use a layered bus architecture to meet the required bandwidth. Figure 3-9 on page 46 shows the AMBA Multi-layer AHB System Interface.

---

1. Originally published in the *CoreConnect Bus Architecture* product brief, 1999. Reprinted with permission of IBM Corporation.

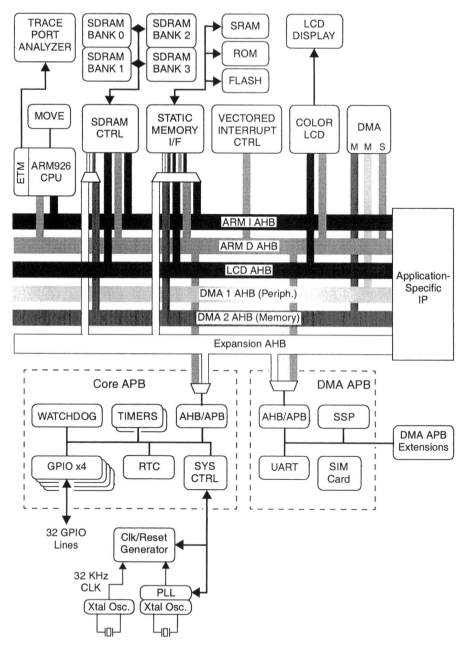

Figure 3-9   Multi-layer AHB System Interface block diagram[1]

---

1. Originally published in the *ARM PrimeXsys Wireless Platform Detailed White Paper, Rev. 4.0*, 2001. Reprinted with the permission of ARM Ltd.

Despite differences in details, there are certain key characteristics shared by all well-designed on-chip bus protocols.

3.5.2   Tristate vs. Mux Buses

When bus structures first migrated from boards onto chips, there was some controversy over whether to use a tristate bus or a multiplexer-based bus. Tristate buses are popular for board-level design, because they reduce the number of wires in the design. However, tristate buses are problematic for on-chip interconnect. It is essential to ensure that only one driver is active on the bus at any one time; any bus contention, with multiple drivers active at the same time, can reduce the reliability of the chip significantly. For high-performance buses, where we want to drive the bus on nearly every cycle, this requirement can produce very timing-critical, technology-dependent designs. Similarly, tristate buses must never be allowed to float; if they float to threshold voltage, they can cause high currents in the receiver, again reducing long-term chip reliability. Either some form of bus-keeper, or a guarantee that exactly one driver is driving the bus at all times, is required. This requirement is particularly difficult to meet during power-on.

Rule – For these reasons, we consider multiplexer-based buses to be the only acceptable architecture for on-chip interconnect. Tristate buses should never be used on chip.

3.5.3   Synchronous Design of Buses

Besides meeting performance goals, on-chip buses must provide a well-defined environment for integrating reusable blocks from a variety of sources. Since our guidelines for reusable designs require a synchronous design style employing single-edge clocking and flip-flops, buses that comply with the same design style will facilitate integration.

Rule – On-chip buses must use a single-clock-edge, flip-flop based architecture.

3.5.4   Summary

In addition to meeting the two rule above, the commercial buses mentioned above all share some other characteristics of good on-chip bus design:

- Separate address and data buses
- Separate control bus(es) that define the kind of transaction taking place
- Support for multiple masters

3.5.5 IP-to-IP Interfaces

Another challenge for early reuse-based designs was the proliferation of point-to-point connections between blocks. For example, in designs such as in Figure 3-10, designers decided that the primary path for data to pass from the I/O block to the data processing block should be a direct, point-to-point connection optimized for maximum performance.

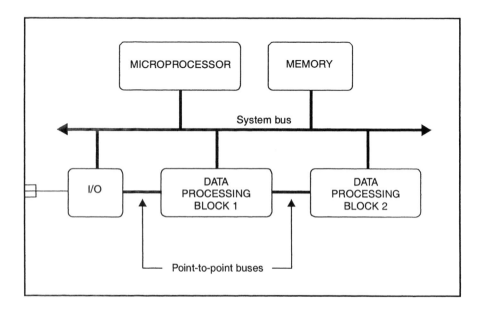

Figure 3-10   Point-to-point connections for system blocks

This approach presents real problems for reuse-based designs. If we want to reuse Data Processing Block 1, but need a different I/O block, say, for a new standard, then we have limited options. We can get an IP from a third-party, but then we will need to develop an interface between the new I/O block and the processing block. Or, we can design the new I/O block ourselves. Either way, the design team will incur unnecessary delays in the project.

To avoid these problems, SoC designers now take a very disciplined approach to all on-chip connections. Wherever possible, data is passed between blocks via an on-chip bus, as shown in Figure 3-11 on page 49. If greater bandwidth is needed, a layered approach like that in Figure 3-9 on page 46 can be used.

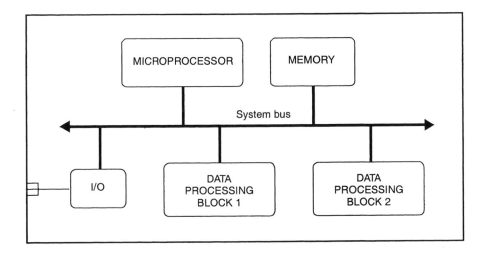

Figure 3-11 Figure 3-10 without point-to-point connections

One instance where point-to-point is appropriate is in the case of an analog block such as a Physical Layer Interface (PHY), as shown in Figure 3-12.

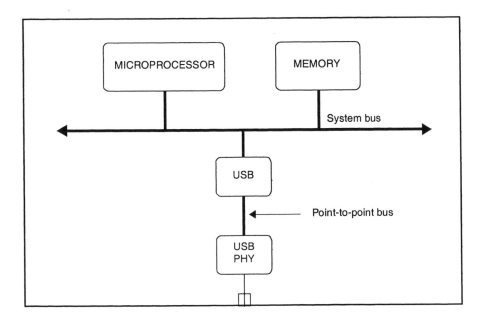

Figure 3-12 Point-to-point connection for a PHY

USB 2.0 can operate at up to 480 Mb/sec., too fast for a synthesized design. A full-custom serializer/deserializer (SERDES) is required to convert the 480 Mb/s serial interface to a 8- or 16-bit parallel interface running at 60 or 30 MHz. Often this SER-DES (also called a PHY) will come from a different design team or third-party vendor than the USB core itself. This presents the classic reuse problem—how to interface the two. The best solution is to have both the PHY and the core use a standard inter-face, as shown in Figure 3-13, so that no additional interfacing design is required from the integrator. UTMI has become a standard for the USB 2.0 PHY/core inter-face, allowing a number of different PHYs and cores to interoperate.

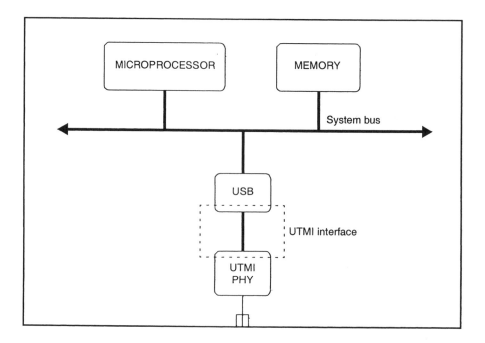

Figure 3-13   UTMI standard interface for USB 2.0 PHY

In summary, we recommend the following guidelines to SoC design teams.

Guideline – The design team should choose an industry-standard bus for the chip design. Wherever possible, blocks should be connected to the bus and/or to the chip's I/O pins. Point-to-point connections between blocks should be used only where abso-lutely necessary, and even then should use an industry-standard interface. Custom interfaces should be used only when no standard exists.

Guideline – In selecting macros for the chip design, either from internal sources or from third-party IP providers, the design team should only accept macros that inter-face directly to the selected bus.

3.6   Design for Bring-Up and Debug: On-Chip Debug Structures

One way to judge the maturity of a design team is the foresight it shows in planning how to bring up first silicon. Inexperienced teams tend to focus exclusively on design and assume that the chip will work. Experienced teams assume that the chip won't work when first powered up and plan accordingly.

Rule – The design team must develop a strategy for the bring-up and debug of the SoC design at the beginning of the design process. The most effective debug strategies usually require specific features to be designed into the chip. Adding debug features early in the design cycle greatly reduces the incremental cost of these features, in terms of design effort and schedule. Adding debug features after the basic functionality is designed can be difficult or impossible, and very time consuming. However, without effective debug structures, even the simplest of bugs can be very difficult to troubleshoot on a large SoC design.

Guideline – Controllability and observability are the keys to an easy debug process.

* *Controllability* is best achieved by design features in the macros themselves. The system should be designed so that each macro can be effectively turned off, turned on, or put into a debug mode where only its most basic functions are operational. This can be done either from an on-chip microprocessor or microcontroller, or from the chip's test controller.
* *Observability* is a major problem for SoC designs. We'd like to be able to put logic analyzer probes on internal nodes in the chip, and debug the chip the way we debug boards. For SoC designs, we can come close to this by adding additional circuitry on the chip to aid observability. We can add circuitry to monitor buses to check data transactions and detect illegal transactions. Another useful approach is to provide a mechanism for observing the internal bus(es) on the chip's I/O pins. Since I/O pins are usually at a premium, this is often done by muxing the bus onto existing I/O pins.

For a general discussion of on-chip debug techniques, see [12]. For a description of ARM's approach to on-chip debug, see [13]. Motorola, Hitachi, Agilent, and Bosch Etas have formed the Nexus 5001 Forum (previously called Nexus Global Embedded Processor Debug Interface Standard Consortium) to come up with a debug interface standard [14,15].

3.7   Design for Low Power

With portable devices becoming one of the fastest growing segments in the electronics market, low-power design has become increasingly important. Traditionally, design teams have used full-custom design to achieve low power, but this approach does not give the technology portability required for reuse-based design. In this section we discuss techniques that result in both low-power and reusable designs.

The power in a CMOS circuit consists of static and dynamic power.

Static Power

For standard cell designs, static current historically has been inherently low, primarily a function of the process technology and the library, rather than the design. Good standard cell libraries have avoided dynamic cells, so that fully static designs, with essentially zero static power, could be implemented.

This situation is changing with today's low-voltage processes. To maintain high performance, library providers are offering low V_t libraries as well as high V_t libraries. The low V_t libraries offer significantly faster cells but at a cost of significant static power. Some aggressive chip designs are using a combination of low V_t and high V_t cells in the same chip design to obtain an optimum balance between high performance and low power.

Achieving very low power designs with low V_t libraries is a big challenge. One technique that is getting a lot of attention is to shut down the power supply to blocks on the chip that use low V_t cells. This technique is just beginning to be investigated.

Dynamic Power

Techniques for lowering the dynamic power of an SoC design are relatively mature, and will be the focus of the rest of this section. The dynamic power of a CMOS design can be expressed as:

$$P = \sum \alpha f C V^2$$

where the sum is over all nodes, and

- α is the switching activity for the node
- f is the clock frequency
- C is the capacitance of the node
- V is the supply voltage

The basic approach to low-power design is to minimize α, C, and V; f is then fixed by the required system performance.

There are two scenarios where power-reducing techniques are applied:

- High-performance designs where low power is a secondary, but still important design criteria (for example, the CPU chip for a laptop computer)
- Extremely low-power designs where low power is the primary objective of the design (for example, battery-operated devices)

Most of the of the power-reduction techniques described below are used for both of these scenarios. One major difference between these two scenarios, though, is in the semiconductor process and standard cell libraries used. For high-performance designs requiring low power, designers are likely to use faster, smaller geometry processes that require a lower voltage to achieve the same performance level. They typically will accept some static power, and will not push the supply voltages to the absolute minimum. Extremely low power designs are likely to use processes and libraries that allow very low voltage operation without significant static currents, and architect the design to use the speed performance that results.

The details of applying power-reduction techniques to these scenarios is are given in the next sections.

3.7.1 Lowering the Supply Voltage

Lowering the supply voltage has the largest effect on power, since power varies with the square of the voltage. Silicon providers have been lowering the standard supply voltage with each new process from 0.5 µm onwards. Running the core of the chip at the lowest possible voltage (consistent with correct functionality) is first step in achieving a very low power design.

Unfortunately, lowering the supply voltage has several adverse effects which must be overcome in other areas of design.

The primary problem with lowering the supply voltage is that it slows the timing performance of the chip. To compensate for this factor, designers typically use pipelining and parallelism to increase the inherent performance of the design. Although this increases area of the design, and thus the overall capacitance, the end result can lower power significantly [16].

I/O voltages must meet the requirements of the board design, and are usually higher (3.3v to 5v) than the minimum voltage that the process will support. Most designers run the I/O at the required voltage, and use a separate, lower voltage power supply for the core logic of the chip.

3.7.2   Reducing Capacitance and Switching Activity

Once we have lowered the supply voltage to the minimum, we need to reduce the capacitance and switching activity of the circuit.

The standard cell library provider can use a variety of techniques to produce a low-power library. The detailed techniques are beyond the scope of this book, but are discussed in [17].

Once we have selected a good low-power library, we can use architectural and design techniques to reduce system power. In real chips, memory design, I/O cells, and the clocking network often dominate overall power. These areas deserve special attention when doing low-power design.

Reducing power in I/O requires minimizing the internal, short-circuit switching current (by selecting the right I/O cell from the library) and minimizing the capacitance of the external load.

Memory Architecture

Reducing power in the on-chip memories again involves both circuit and architectural techniques. Most silicon providers have memory compilers that can produce a variety of memory designs that trade off area, power, and speed.

The memory architecture itself can reduce power significantly. Instead of using a single, deep memory, it may be possible to partition the memory into several blocks, selected by a decode of the upper or lower address bits. Only the block being accessed is powered up. This approach again produces redundant logic (in extra decode logic), so it reduces power at the expense of (slightly) increasing area. This technique is shown in Figure 3-14 and described in more detail in [16], where an 8x reduction in RAM power was achieved.

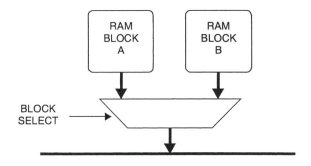

Figure 3-14   Multi-block RAM architecture

Clock Distribution

In pipelined designs, a significant portion of the overall power is in the clock, so reducing power in the clock distribution network is important. As few different clocks as possible should be used. Single clock, flop-based designs can reduce power by 50% over latch-based dual, non-overlapping clock designs.

Shutting down clock distribution to part of the circuit by clock gating can significantly reduce chip power. Clock gating, however, can be very technology dependent; careful design is required to assure a portable, reusable design.

There are two basic types of clock gating: gating the clock to a block of logic, or gating the clock to a single flop.

In Figure 3-15, a central clock module provides separate gated clocks to Block A and Block B. Significant power savings are realized because whole blocks can be shut down when not being used, and because the entire clock distribution to the block is disabled. Since large buffers are often used in clock distribution networks, just shutting down the clock inputs to these buffers can result in significant savings.

The actual clock gating circuit itself can be non-trivial. Disabling the clock in such a way as to avoid generating a glitch on the clock line requires careful design, and a detailed knowledge of the timing of the gates used. For this reason, the clock gating circuit itself tends to be technology dependent and not reusable.

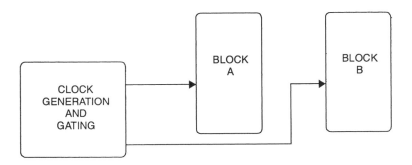

Figure 3-15   Block-level clock gating

Isolating the clock gating in a separate clock generation block allows Block A and Block B to be designed to be completely reusable. The clock generation block can be made small, so that its technology-dependent design can be manually verified for correctness.

In some cases, it may not be possible to gate the clock to an entire block, and the designer may want to gate the clock on a flop-by-flop basis. This case usually occurs on flops where we selectively hold data, as shown in Figure 3-16.

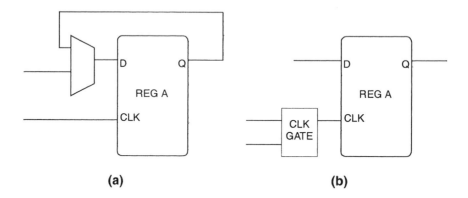

(a)   (b)

Figure 3-16   Clock gating at individual flip-flops

In Figure 3-16a, Register A has its data selectively held by the mux. Figure 3-16b shows the equivalent circuit using clock gating instead, which results in lower power.

The implementation Figure 3-16a is generated directly from the RTL, while the implementation in Figure 3-16b is generated by a power synthesis tool.

Guideline – The approach in Figure 3-16b is not recommended for reusable designs, since the clock gating function is inherently technology dependent. Rather, we recommend designing as in Figure 3-16a. Today's advanced power synthesis tools can detect this configuration, and, working in conjunction with clock tree physical design tools, automatically convert it to the configuration in Figure 3-16b.

By designing and coding the circuit without clock gating, engineers can assure that the design is technology independent and reusable.

3.7.3   Sizing and Other Synthesis Techniques

The next major technique for reducing chip power involves optimizing the gate-level design for low power.

Gate sizing can produce a significant power savings in many designs. This technique consists of reducing the drive strength of gates to the lowest level that meets the timing requirements for the design. Synthesis tools can do this automatically, without any requirement for changing the RTL code.

Some incremental improvement can be gained by restructuring logic to reduce the number of intermediate, spurious transitions in the logic. Again, synthesis tools can do this automatically.

3.7.4   Summary

In [16] the results of several low-power chip designs are reported. The results for show:

- A 21x reduction in power by lowering the voltage from 5v to 1.1v. Note that this reduction involved moving from a standard, large geometry process to a smaller, lower voltage process specifically designed for low power. A number of architectural changes had to be made to maintain performance at this low voltage.

- A 3-4x reduction from gate sizing, low-power I/O cells, and similar gate-level optimizations. Note that the results reported are from manual designs. Today's synthesis tools include gate sizing and other gate-level optimizations. Any manual optimizations post-synthesis are not likely to result in significant power savings, except in the selection of I/O cells.

- A 2-3x improvement by clock gating.

- An 8x improvement in a memory array by using the multi-block technique described above.

Thus, with a good low-power library, low-power design for reuse is possible through a combination of architectural techniques and the proper use of power synthesis tools. These techniques can produce designs that are fully reusable and are quite close to full-custom designs in power consumption. Considering that overall chip power is likely to be dominated by I/O and memory, the small increase in power from the logic in the chip is more than offset by the time-to-market advantage of having reusable blocks.

3.8   Design for Test: Manufacturing Test Strategies

Manufacturing test strategies must be established at specification time. The optimal strategy for an individual block depends on the type of block.

3.8.1   System-Level Test Issues

Rule – The system-level chip manufacturing test strategy must be documented.

Guideline – On-chip test structures are recommended for all blocks. It is not feasible to develop parallel test vectors for chips consisting of over a million gates. Different kinds of blocks will have different test strategies; at the top level, a master test controller is required to control and sequence these independent test structures.

Guideline – Include an asynchronous set or reset function for all flip-flops. This allows the chip to be put in a known state without requiring a complex initialization routine. This, in turn, makes test programs easier to create.

3.8.2   Memory Test

Guideline – Some form of BIST is recommended for RAMs, because this provides a rapid, easy-to-control test methodology. However, some BIST solutions are not sufficient to test data retention.

3.8.3   Microprocessor Test

Microprocessors targeted for use in SoC designs are increasingly using full-scan techniques. The key challenge in integrating the processor into the overall scan testing of the chip involves dealing with the interface between the processor and the rest of the chip. Figure 3-17 on page 59 shows how shadow registers can be used to overcome this problem. Without these registers, the logic between the processor's scan flops and the scan flops of the rest of the chip is very hard to test. The scan test patterns for the processor would depend on the scan patterns of the rest of the chip, and vice versa. This problem quickly becomes intractable. The shadow registers provide isolation between the processor scan chain and that of the rest of the chip, making scan test pattern generation two separate, local problems. The shadow registers at the processor inputs allow the scan chain to drive inputs to the processor and to sample the values that the external circuitry applies to the chip inputs. In this sense, the shadow registers function both as part of the processor's scan testing and as part of the scan testing of the rest of the chip. The latest test tools can automatically insert shadow registers and multiplexers for scan chains.

Guideline – Use shadow registers to facilitate full-scan testing with embedded processors.

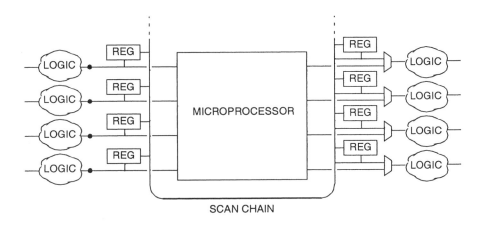

Figure 3-17 Shadow registers for full-scan test logic

3.8.4 Other Macros

Guideline – For most other blocks, the best choice is a full-scan technique. Full scan provides very high coverage for very little design effort. The chip-level test controller needs to manage the issue of how many scan chains are operated simultaneously, and how to connect them to the chip-level I/O.

3.8.5 Logic BIST

Logic BIST is a variation on the full-scan approach. Where full scan must have its scan chain integrated into the chip's overall scan chain(s), logic BIST uses a Linear Feedback Shift Register (LFSR) to generate the test patterns locally. A signature recognition circuit checks the results of the scan test to verify correct behavior of the circuit.

Logic BIST has the advantage of keeping all pattern generation and checking within the macro. This provides some element of additional security against reverse engineering of the macro. It also reduces the requirements for scan memory in the tester and allows testing at higher clock rates than can be achieved on most testers. Logic BIST does require some additional design effort and some increase in die area for the generator and checker. Tools to automate this process are becoming available.

Logic BIST is currently being used in some designs, but it is much less common than standard full-scan testing. The success of logic BIST in the long term probably depends on the ability of test equipment manufactures to keep up with the need for ever-increasing scan memory in the tester. If the test equipment fails to provide for scan test of large chips, logic BIST will become the test methodology of choice for SoC designs.

3.9   Prerequisites for Reuse

We conclude this chapter with a discussion of the technical infrastructure that must be in place for the standard model of reuse to be successful.

3.9.1   Libraries

First of all, design teams must have access to high-quality standard cell libraries. These libraries should provide a full set of views, including synthesis, physical, and power views. These libraries need to be validated in hardware so that design teams can have a high degree of confidence in their timing and power characteristics. Finally, the libraries should have a reasonably accurate, validated wire load models to enable useful pre-layout synthesis and timing analysis of designs. Libraries should not contain any dynamic cells.

These libraries need to be tested in the SoC flow before they can be considered completely validated. A number of subtle problems, such as not modeling antenna rules correctly or using non-standard definitions for rise times, can bring a large chip design project to a screeching halt. Testing the libraries through the entire flow can help prevent significant delays on later projects.

These libraries should be available as early as possible. In some semiconductor companies, libraries are not available to design teams until after the process is on line. This is too late; many designs are started while the new process is being developed. In some cases design teams have designed their own libraries to allow design work to proceed. This practice can lead to the proliferation of unvalidated, high-defect libraries.

These libraries should also include a set of memory compilers. These memory compilers should provide for some tradeoffs between power, area, and timing performance. They should support single and multiple port configurations, and provide fully synchronous interfaces. (Generating a write pulse in standard cell logic requires technology-dependent, non-reusable design practices.)

If the target technology supports flash EEPROM and/or DRAM, then the memory compilers should support these as well.

Libraries should also provide special cells for clock gating.

Although not always considered part of the library, certain analog blocks occur so often in chip designs that they should be provided along with the library. These include Phase-Locked Loop (PLL) clock generators and basic analog-to-digital and digital-to-analog converters. PLLs in particular are very demanding designs, and it makes no sense to force individual design teams to develop their own.

As chips become larger, and reuse of multiple hard blocks from multiple sources becomes more common, it becomes increasingly likely that an SoC design will use multiple libraries. Also, as mentioned in the section on low power, designers are beginning to do chips that use multiple libraries (low Vt and high Vt) to achieve the best balance between performance and power. These libraries must be designed to interoperate to avoid problems during physical design.

Interoperabilty problems usually show up during DRC and LVS. Name conflicts are the most common problem.

Guideline – Cell names should be unique to the library.

Another problem for combining hard IP from multiple sources is the question of what DRC rules and LVS decks should be used during physical verification. Hard IP providers should specify the DRC rules and LVS decks to be used as part of the IP deliverables.

3.9.2   Physical Design Rules

One common problem in large designs is that several pieces of hard IP are integrated from different sources. For instance, an automotive group may use a processor from a computer division and a DSP from a wireless division. If these blocks have been designed with different physical design rules, and verified using different DRC decks, then physical verification at the chip level can be a major problem. The design team will be hard pressed to find or develop a DRC deck that will work for both blocks.

We strongly recommend that, for a given process, standard DRC and LVS decks be developed and validated. These decks should be used by all design teams, so that physical designs (hard IP) can be exchanged and integrated without undue effort.

References

1. Chinnery, David G. and Keutzer, Kurt. *Closing the Gap Between ASIC and Custom*. Kluwer, 2002.

2. Chinnery, D. G., Nikolic, K., and Keutzer, K. "Achieving 550 MHz in an ASIC Methodology," *Proceedings of the 38th Design Automation Conference, 2001*, pp. 420-425.
   Website: http://www.dac.com/39th/archives.html

3. Chinnery, D. G. and Keutzer, K. "Closing the Gap Between ASIC and Custom: An ASIC Perspective," *Proceedings of the 37th Design Automation Conference, 2000*, pp. 637-642.
   Website: http://www.dac.com/39th/archives.html

4. Sylvester, Dennis and Keutzer, Kurt. "Getting to the Bottom of Deep Submicron," *Proceedings of ICCAD, 1998*. pp. 203-211.
   Website: http://www.sigda.org/Archives/ProceedingArchives/Iccad/Iccad98

5. Chaney, Thomas, "Measured Flip-Flop Responses to Marginal Triggering," *IEEE Transactions of Computers*, Volume C-32, No. 12, December 1983, pp. 1207-1209.

6. Horstmann, Jens U., Hans W Eichel, and Robert L. Coates, "Metastability Behavior of CMOS ASIC flip-flops in theory and test," *IEEE Journal of Solid-State Circuits*, Volume 24, No. 1, February 1989, pp. 146-157.

7. Poulin, Jeffrey. *Measuring Software Reuse: Principles, Practices, and Economic Models*. Addison-Wesley, 1996.

8. Jones, Capers. *Applied Software Measurement: Assuring Productivity and Quality*. McGraw-Hill, 1996.

9. ARM AMBAbus website: http://www.arm.com/arm/AMBA?OpenDocument

10. MIPS BusBridge website: http://www.mips.com/publications/bus_bridge.html

11. IBM CoreConnect bus website: http://www-3.ibm.com/chips/techlib/techlib.nsf/productfamilies /CoreConnect_Bus_Architecture

12. Neugass, Henry. "Approaches to on-chip debugging," *Computer Design*, December 1998.

13. Goudge, Liam. *Debugging Embedded Systems*. White Paper, ARM, Ltd.
    Website: http://www.arm.com/Documentation/WhitePapers/DebugEmbSys

14. Cole, Bernard. "User demands shake up CPU debug traditions," *EE Times*.
    Website: http://www.eetimes.com/story/OEG19990216S0009

15. Nexus website: http://www.nexus5001.org/standard.html

16. Chandrakasan, Anantha and Brodersen, Robert. *Low Power Digital CMOS Design*. Kluwer Academic Publishers, 1995.

17. Chandrakasan, Anantha (editor) and Brodersen, Robert (editor). *Low Power CMOS Design*. IEEE, 1998.

CHAPTER 4

The Macro Design Process

This chapter addresses the issues encountered in designing hard and soft macros for reuse. The topics include:

- An overview of the macro design workflow
- Contents of a design specification
- Top-level macro design and partitioning into subblocks
- Designing subblocks
- Integrating subblocks and macro verification
- Productization and prototyping issues

4.1 Overview of IP Design

One key to successful SoC design is to have a library of reusable components from which to build the design. This chapter describes the process of designing reusable components, or IP.

4.1.1   Characteristics of Good IP

To support the broadest range of applications, and provide the highest reuse benefits, IP should have these features:

- Configurable to meet the requirements of many different designs
- Standard interfaces
- Compliance with defensive design practices to facilitate timing closure and functional correctness
- Complete set of deliverables to facilitate integration into a chip design

Configurability

Most IP has to be configurable to meet the needs of many different designs (and if it doesn't meet the needs of many different designs, it is not worth making the investment required to make it reusable). For example:

- Processors may offer different implementations of multipliers, caches, and cache controllers.
- Interface blocks like USB may support multiple configurations (low-speed, full-speed, high-speed) and multiple interfaces for different physical layer interfaces.
- Buses and peripherals may support configurable address and data bus widths, arbitration schemes, and interrupt capability.

Configurability is key to the usability of IP, but also poses great challenges, since it makes the core harder to verify.

Standard Interfaces

As mentioned in Chapter 3, reusable IP should, wherever possible, adopt industry-standard interfaces rather than unique or core-specific interfaces. This makes it possible to integrate many different cores without having to build custom interfaces between the IP and the rest of the chip.

Compliance to Defensive Design Practices

Chapter 5 provides details of design and coding of IP for timing closure, ease of verification, and ease of packaging for reuse. These are based on the more general requirements for good design outlined in Chapter 3. All of these, ultimately, come down to the age-old principle of good engineering: keep the design as simple as possible.

Complete Set of Deliverables

- Synthesizable RTL (encrypted or unencrypted)
- Verification IP for verifying the core stand-alone and for chip-level verification
- Synthesis scripts
- Documentation

Deliverables are discussed in detail in Chapter 9.

4.1.2   Implementation and Verification IP

As reuse has matured, Verification IP (VIP)—such as Bus Functional Models (BFMs), monitors, and test suites—have become more important. Figure 4-1 on page 66 shows how a USB bus functional model and bus monitor, as well as a model of a DRAM, can be used to facilitate verification of an SoC (the Implementation IP, or IIP). Figure 4-2 on page 67 shows how the same USB BFM and monitor can be used in a verification testbench by the macro design team to verify the USB macrocell. To allow SoC designers to use implementation IP effectively, the IP developer must provide VIP as well as the IIP.

The VIP provides a high-level mechanism for generating transactions on the interfaces of the IIP. This is useful to:

- Do initial verification of the RTL
- Allow rapid configuration of the testbench for testing different configurations of the IP
- Ship with the IP to allow the integrator to test his configuration of the IIP before integrating it into the chip
- Integrate into the chip-level test bench

The difficulty in verifying configurable IP and the even greater difficulty of verifying SoCs mandates the development of VIP. This kind of high-level, transaction-based, disciplined approach to verification is the only way to achieve the required robustness in SoC and IIP.

VIP is discussed more in Chapter 7.

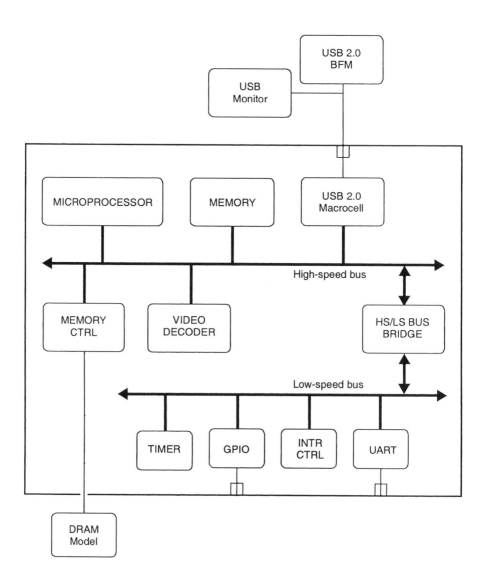

Figure 4-1   SoC verification scheme

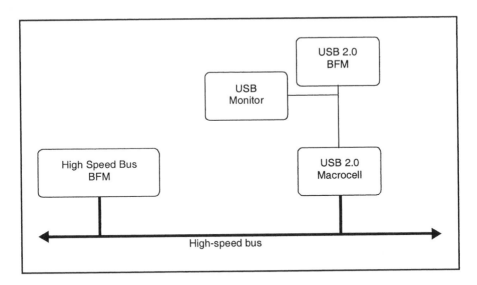

Figure 4-2   USB macro testbench (portion of Figure 4-1)

4.1.3   Overview of Design Process

Figure 4-3 summarizes the first five major phases of IP design.

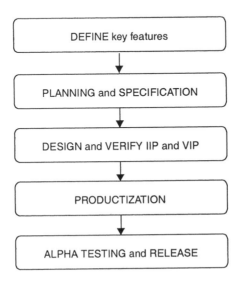

Figure 4-3   Phases of IP design

The IP design phases shown in Figure 4-3 can be described as follows:

1. **Define Key Features** – The first step in the macro design process is to define the key features of the macro. These features include basic functionality and the configuration parameters that will allow the macro to be used in the various target applications. Defining the key features requires a complete understanding of the application and the usage model for the macro.

2. **Planning and Specification** – The second step in the macro design process is to develop a detailed functional specification for the macro, a detailed specification for the verification IP, and a detailed plan for the rest of the project.

3. **Design and Verification of the IIP and VIP** – Once the specifications and plan have been defined, the team moves to the detailed design and verification of the macro.

4. **Productization** – Once the design has been completed, additional testing is performed, and the final deliverables, including user documentation, are packaged in a form that can be delivered to the end user.

5. **Alpha Testing and Release** – The final deliverables are then tested to make sure they are complete and ready to be used by the SoC designer.

This items listed above are the major tasks in the macro design process. For more information, see the discussion of the spiral workflow on page 11 and Figure 4-6 on page 77.

It is often said that the earlier a mistake is made in the design process, the more costly it is. This is certainly true of designing IP. Making a mistake in defining a key feature, and only discovering it once the IP is shipped, is enormously expensive, since a large amount of work must be redone. Similarly, having specifications that are poor or incomplete, and discovering this only during integration and test, or worse, when the customer starts using the product, can again result in a huge amount of redo effort. Thus, steps listed in Figure 4-3 are not only in order of sequence, but in order of importance as well.

4.2   Key Features

The key features for an IP are basically an initial specification, defining the key functionality required for the IP. For standards-based IP (USB 2.0, IEEE 1394) this includes which version of the standard will be implemented, and what features from the standard are supported and which are not supported. Being able to configure an IP can be critical to enabling its use in a variety of applications. The key features should define the configuration parameters for the IP.

Key features are typically captured in a brief document or a set of slides.

4.3   Planning and Specification

During the planning and specification phase of the project, the team develops the key documents that will steer the project through the rest of its life-cycle.

The key documents developed during the planning and specification phase typically include:

- Functional specification
- Verification specification
- Packaging specification
- Development plan

4.3.1   Functional Specification

The functional specification provides a complete description of the functionality of the design from the users perspective—that is, from the perspective of the SoC designer who is integrating the design into a chip. Once this functionality is defined, the design of the IP becomes a local and autonomous activity; no additional information is required and no additional, external dependencies exist.

A specification for a macro consists of:

- Pin definitions
- Parameter definitions
- Register definitions
- Additional information to explain how the pins, parameters, and registers work
- Performance and physical implementation requirements

The sections of a functional specification would include the following sections:

Overview
This section provides a brief, high-level overview of the functionality of the design. In particular, if the design needs to comply with a specific standard, such as an IEEE standard, the standard must be specified here. It can also be useful to describe some example applications here.

Functional Description
This section provides a more detailed overview of the functionality of the design. One or more block diagrams, along with some brief text, can help describe the functionality and the internal structure of the design. Although, strictly speaking, the specification describes the externally visible functionality of the design, this is often most easily understood by understanding the component parts of the design and their individual functionality. This is approach is just another application of the concept of locality. We break the

difficult task of understanding a complex IP into a set of smaller, local tasks of understanding its subblocks, and then understanding their interaction.

For macros that implement complex algorithms, it may be helpful to describe those algorithms in C or pseudocode.

Pin Definitions

This section describes in detail the most basic interface to the macro—the I/O pins. Each pin (or bus) is defined, including its name, direction (input, output, or inout), polarity (active high or active low), registered or not, width (for buses), and a brief description of its function.

Parameter Definitions

This section describes the parameters that the user can set to select a specific configuration of the IP. Each parameter is defined, including it name, legal values, default value, and a description of its function. In addition, any dependency of the parameter's value on other parameters is described. For instance, if the parameter OUTPUT_DATA_WIDTH must be less than or equal to INPUT_DATA_WIDTH, then this dependency should be noted.

Register Definitions

This section describes the software-programmable registers in the design. Each register is defined, including it name, address (or address offset from some base address), type (read, write, or read/write), width, reset value, and a description of its function.

Additional Information

Strictly speaking, the function of the design is completely defined by the function of the pins, parameters, and registers. But the sections described above are necessarily terse, and often require additional explanation. Pin descriptions need to be supplemented with descriptions of how the pins interact to create transactions—typically register or memory reads and writes.These transactions are often best described by waveforms. Additional drawings and block diagrams can help clarify how transactions on the pin interfaces and the stored state in the registers interact to produce the functionality of the IP. Text can also be used to clarify the functionality of the design, but text should be used as a last resort. Text is remarkably poor at describing complex technical concepts.

Performance and Physical Implementation Requirements

This section describes the clock and reset requirements for the design—target clock frequencies, number of clock domains, number of resets and their timing relationships. Any special requirements for meeting power or performance goals should be documented here.

4.3.2   Verification Specification

The verification specification defines the test environment that is used to verify the IP, including any bus functional models and monitors that must be developed or purchased to build the environment. It also describes the approach to verifying the IP—what will be tested by directed test, what random testing will be done, and what metrics will be used to determine that it is ready to ship.

The verification specification is described in more detail in Chapter 7.

4.3.3   Packaging Specification

The packaging specification defines any special scripts that are provided as part of the final deliverables. These scripts typically include installation scripts, configuration scripts, and synthesis scripts. For hardened IP, this specification also lists additional views that must be created.

4.3.4   Development Plan

The functional, verification, and packaging specifications define the technical contents of the project. The development plan describes how these contents will be produced. The development plan includes:

Deliverables
This section describes the deliverables for the project: what files and documents will be created, archived, and delivered at the end of the project.

Schedule
This section includes key milestones and dates for the project.

Resource Plan
This includes the people, tools, software, computer, and network requirements for the project.

Exceptions to Coding and Design Guidelines
This could include, for instance, exceptions to the *RMM* coding guidelines.

Supported Environments
This section documents what versions of what tools and libraries are supported, including what languages (Verilog, VHDL, etc.), what simulators, and what standard cell libraries. This lists forms the testing matrix for the final product—what configurations of the macro will be tested with what versions of what simulators. It also defines what configurations of the macro will be synthesized with which version of the synthesis tools and with which standard cell libraries.

Support Plan

Who will support the product, and what is the anticipated workload for support. Also, what application notes will be developed to help reduce the support load.

Documentation Plan

What user documents will be developed, by whom, and when.

Licensing Plan

Will the macrocell be licensed, and, if so, how.

Release Plan

This section describes the release process for the final macro. This usually includes some form of alpha testing before release, and then some form of beta or limited production period before the final, full-production release.

4.3.5   High-Level Models as Executable Specifications

For many standards-based IPs, developing the functional specification is a reasonably straightforward process. To a great degree, the functionality and interfaces are defined by the standard itself. For many other designs, it may be useful to develop a high-level model as a mechanism for exploring alternative algorithms or architectures. This high-level model can then act as a executable functional specification for the design, complementing the written functional specification.

Developing a high-level model also allows the development of testbenches and test suites early in the project. For macros that have software content, the behavioral model provides a high-speed simulation model early in the design cycle. The software developers can use this model for software design and debug while the detailed design is done. This approach can be essential for meeting time-to-market goals.

High-level models are developed as part of the functional specification phase of the project, before detailed design is started. Decisions made during high-level modeling can dramatically affect the detailed design.

High-level models can be written at the algorithmic or transaction level. Algorithmic models are purely behavioral, with no timing information. They can be particularly useful for designs in multimedia or wireless, for instance, when fundamental algorithmic decisions need to be made affecting bandwidth requirements, signal-to-noise performance, and compression ratios.

Transaction-level models are cycle-accurate models that model transactions at their interfaces as atomic events, rather than as a series of event-level activities on pins. By abstracting out the pin-level behavior of the design, transaction-level models can be quite accurate but still much faster than RTL-level models. This speed-accuracy tradeoff is what makes transaction-level models so useful in evaluating multiple architectures for a design.

4.4   Macro Design and Verification

Figure 4-4 shows the macro design process up to the point of integrating subblocks back into the parent macro. Then Figure 4-5 on page 74 shows the process of integrating the subblocks.

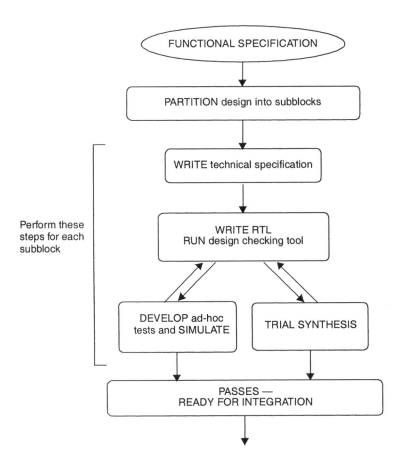

Figure 4-4   Subblock development process

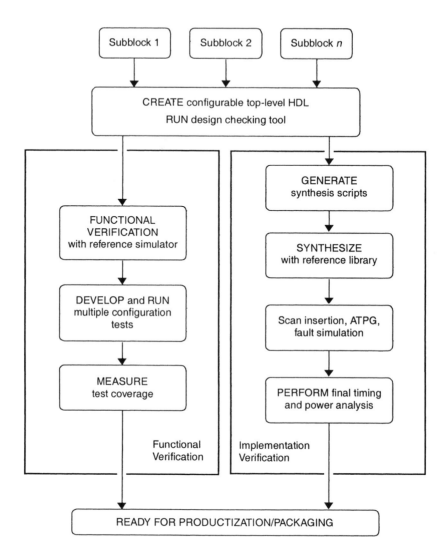

Figure 4-5   Subblock integration process

The major steps in the macro design processes shown in Figure 4-4 and Figure 4-5 are:

1. **(Macrocell) Technical Specification** – For all but the simplest macros, the functional specification does not contain enough information to allow coding to begin. For reasonably complex macros, technical specifications are needed. Once the functional specification is developed and approved, the team needs to develop the technical specification for the macrocell. While the functional specification

describes the design as seen from its interfaces, the *technical specification* describes the internals of the macrocell and captures the details of the design. In particular, it describes the partitioning of the macrocell into subblocks—parts of the design small enough to be designed and coded by a single engineer. Most of the true design work is done in developing the technical specifications, since this is where the internal structure and detailed interfaces are defined. The quality of the technical specification is a key factor in determining the time required for the rest of the design process. A good technical specification allows the designer to code once and to verify quickly. A poorly thought-out specification results in many iterations through the code/test/synthesis loop.

2. **(Subblock) Technical Specifications** – For complex designs, technical specifications for the subblocks are required. Once the partitioning is complete, the subblock designer develops a technical specification for the subblock. This specification includes the timing and functionality of the interfaces to other subblocks, as well as internal structure and critical algorithms. The amount of detail required in these technical specifications is very design-specific. For some simple macrocells, the top-level technical specification may be detailed enough that coding can be done directly from it. For large, complex macrocells, each subblock may require extensive, detailed specifications. But the combination of the top-level technical specification and the subblock technical specifications should provide enough detail that the engineer can code directly from it. That is, all the key design decisions have been made before coding begins. The specifications for all subblocks are reviewed by the team and checked for consistency.

3. **(Subblock) RTL** – Once the technical specifications are written and reviewed, the engineer can develop the RTL. As the code is written, a design checking tool should be run on it to verify that it complies with the team's design and coding guidelines.

4. **(Subblock) Functional Verification** – The designer constructs a simple testbench to verify the functionality of the subblock. One of the key differences between subblock design and macrocell design is that the testbench for the subblock is ad hoc, used for initial testing, and then discarded at the end of the project. Its only purpose is to assure that the subblock is functionally correct before it is integrated into the overall macrocell. Even though it is ad hoc, this subblock verification needs to be rigorous: finding and fixing bugs is much easier at the subblock level than after integration.

5. **(Subblock) Implementation Verification** – The engineer runs synthesis on the subblock to make sure that the subblock compiles correctly and meets initial timing goals.

6. **(Macrocell) RTL** – Once all the subblocks are complete, the team is ready to integrate them into the top-level RTL. The top-level RTL is typically written to be configurable. Port widths, signal polarities, and even functionality may be selected by the end user to select the specific configuration needed for the target design. Scripts may need to be written to generate specific configurations for verification. A design checking tool should be run on the final RTL to make sure the full design complies with design and coding standards.

7. **(Macrocell) Functional Verification** – It is essential to develop a thorough functional test suite and to run it on the final macro design. The design team must run this test on a sufficient set of configurations to ensure that the macro is robust for all possible configurations. The verification strategy for the entire macro is discussed in Chapter 7 of this manual.

8. **(Macrocell) Implementation Verification** – In parallel with the functional verification of the macro, the design team needs to verify that the macrocell can be implemented and meet the timing, area, and testability of the final product.

Develop synthesis scripts

The design team needs to develop a top-level synthesis script. For parameterizable macros, where the number of instances of a particular subblock may vary, this presents a particular challenge. It may be necessary to provide a set of scripts for different configurations of the macro. It may also be useful to provide different scripts for different synthesis goals: one script to achieve optimal timing performance, another to minimize area.

Run synthesis

The design team must run synthesis on a sufficiently large set of configurations to ensure that synthesis will run successfully for all configurations. In general, this means synthesizing both a minimum and maximum configuration. Note that the final synthesis constraints must take into account the fact that scan will later be inserted in the macro, adding some setup time requirements to the flops.

Perform scan insertion

The final RTL code must also meet the testability requirements for the macro. Most macros will use a full-scan test methodology and require 95% coverage (99% preferred).

Use a test insertion tool to perform scan insertion and automatic test pattern generation for the macro. As part of this process, the test insertion tool should also report the actual test coverage for the macro.

After scan insertion, the design team uses a static timing analysis tool to verify the final timing of the macro.

Perform power analysis

If power consumption is an issue, the design team uses a power analysis tool to ensure that power consumption is within specification.

4.4.1 Summary

It is important to note that the separation of the design process into distinct phases does not imply a rigid, top-down design methodology. Frequently, some detailed design work must be done before the specification is complete, just to make sure that the design can be implemented.

A rigid, top-down methodology says that one phase cannot start until the preceding one is completed. We prefer a more mixed methodology, which simply says that one phase cannot complete until the preceding one is completed.

Figure 4-6 shows how the various activities required to develop a reusable macro can proceed in parallel.

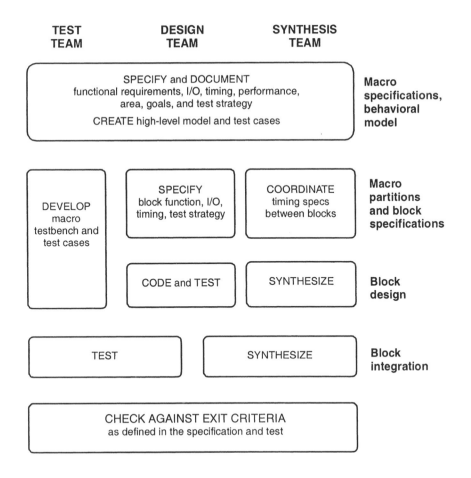

Figure 4-6 Parallel activities in macro development

4.5   Soft Macro Productization

The final phase of macro design consists of productizing the macro, which means creating the remaining deliverables that system integrators will require for reuse of the macro. This chapter describes the productization of soft macros only. The development and productization of hard macros is described in Chapter 8.

4.5.1   Productization Process

The soft macro productization phase, shown in Figure 4-7 on page 79, is complete when the design team has produced and reviewed the following components of the final product:

* Versions of the code, testbenches, and tests that work in both Verilog and VHDL environments
* Supporting scripts for the design

    This includes the installation scripts and synthesis scripts required to build the different configurations of the macro.
* Documentation

    This includes updating all the functional specifications and generating the final user documentation from them.
* Final version locked in a version control system

    All deliverables must be in a revision control system to allow future maintenance. For more information, see "Revision Control Systems" on page 265.

4.5.2   Activities and Tools

The soft macro productization process involves the following activities and tools:

Develop a prototype chip

A prototype chip is essential for verifying both the robustness of the design and the correctness of the original specifications. Some observers estimate that 90% of chips work the first time, but only 50% of chips work correctly in the system.

Developing a chip using the macro and testing it in a real application with real application software allows us to:

* Verify that the design is functionally correct.
* Verify that the design complies with the appropriate standards (for instance, we can take a PCI test chip to the PCI SIG for compliance testing).
* Verify that the design is compatible with the kind of hardware/software environment that other integrators are likely to use.

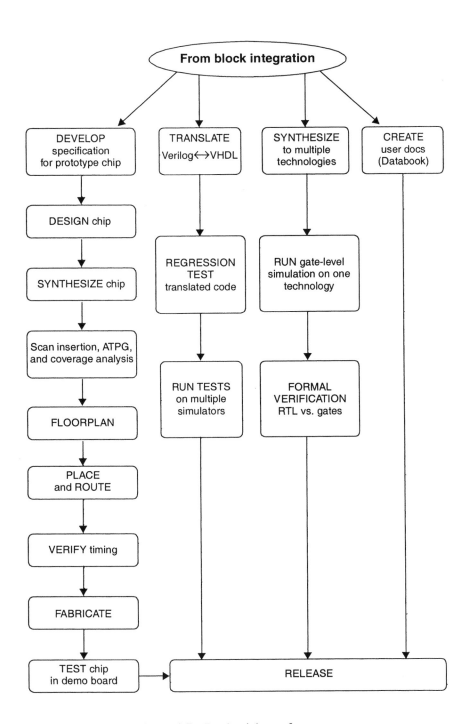

Figure 4-7   Productizing soft macros

The prototype chip can be FPGA or an ASIC, depending on the needs of the project. See Chapter 7 for more discussion on the tradeoffs between these two approaches. If an FPGA is used, the macro should still be taken through a full ASIC place and route process to verify that there are not timing closure or other physical design problems.

The process for developing the prototype chip is a simple ASIC flow appropriate for small chip design. It is assumed that the chip will be a simple application of the macro, perhaps twice the size of the macro itself in gate count.

Provide macro and testbench in both Verilog and VHDL

To be widely useful, the macro and its testbenches must be available in both the Verilog and VHDL languages. Commercial translators are available. These translators do a reasonable job on RTL code but still present some challenge for translating testbenches.

After the code and testbenches have been translated, they must be re-verified to validate the translation.

Test on several simulators

In addition, the macro and testbenches should be run on the most popular simulators in order to ensure portability. This is particularly important for the VHDL simulators, which have significant differences from vendor to vendor.

Synthesize on multiple technologies

The macro should be synthesized using a variety of technologies to ensure portability of the scripts and that the design can meet its timing and area goals with the ASIC libraries that customers are most likely to use.

Perform gate-level simulation

Gate-level simulation must be run on at least one target technology in order to verify the synthesis scripts.

Formal verification

Using formal verification, we can verify that the final netlist is functionally equivalent to the original RTL.

Create/update user documentation

The functional specifications created during the design process are usually not the best vehicle for helping a customer use the macro. A set of user documents must be developed that addresses this need. The components of this documentation are described in Chapter 9 of this manual.

Alpha testing and release

Once productization is complete, the macro is ready for alpha testing. This testing is typically done by someone outside the design team. The purpose of alpha testing is to test the usability of the final deliverables, especially the documentation. Ideally, alpha testing involves designing the macros into a small kit, thus using the macro deliverables exactly as an integrator would. Once alpha testing is complete, the macro is ready for release.

CHAPTER 5   *RTL Coding Guidelines*

This chapter offers a collection of coding rules and guidelines. Following these practices helps to ensure that your HDL code is readable, modifiable, and reusable. Following these coding practices also helps to achieve optimal results in synthesis and simulation.

Topics in this chapter include:

- Basic coding practices
- Coding for portability
- Guidelines for clocks and resets
- Coding for synthesis
- Partitioning for synthesis
- Designing with memories
- Code profiling

5.1   Overview of the Coding Guidelines

The coding guidelines in this chapter are based on a few fundamental principles. The basic underlying goal is to develop RTL code that is simple and regular. Simple and regular structures are inherently easier to design, code, verify, and synthesize than more complex designs. The overall goal for any reusable design should be to keep it as simple as possible and still meet its functional and performance goals.

The coding guidelines detailed in this chapter provide the following general recommendations:

- Use simple constructs, basic types (for VHDL), and simple clocking schemes.
- Use a consistent coding style, consistent naming conventions, and a consistent structure for processes and state machines.
- Use a regular partitioning scheme, with all module outputs registered and with modules roughly of the same size.
- Make the RTL code easy to understand by using comments, meaningful names, and constants or parameters instead of hard-coded numbers.

By following these guidelines, developers should be better able to produce code that converges quickly to the desired performance, in terms of functionality, timing, power, and area.

Design checking tools provide an automated way to check RTL for compliance to design and coding guidelines.

5.2   Basic Coding Practices

The following guidelines address basic coding practices, focusing on lexical conventions and basic RTL constructs.

5.2.1   General Naming Conventions

Rule – Develop a naming convention for the design. Document it and use it consistently throughout the design.

Guideline – Use lowercase letters for all signal names, variable names, and port names.

Guideline – Use uppercase letters for names of constants and user-defined types.

Guideline – Use meaningful names for signals, ports, functions, and parameters. For example, do not use `ra` for a RAM address bus. Instead, use `ram_addr`.

Guideline – If your design uses several parameters, use short but descriptive names. During elaboration, the synthesis tool concatenates the module's name, parameter names, and parameter values to form the design unit name. Thus, lengthy parameter names can cause excessively long design unit names when you elaborate the design with synthesis tools.

Guideline – Use a consistent name for the clock signal, such as clk. If there is more than one clock in the design, use clk as the prefix for all clock signals (for example, clk1, clk2, or clk_interface).

Guideline – Use the same name for all clock signals that are driven from the same source.

Guideline – For active-low signals, end the signal name with an underscore followed by a lowercase character (for example, _b or _n). Use the same lowercase character consistently to indicate active-low signals throughout the design.

Guideline – Use a consistent name for reset signals, such as rst. If the reset signal is active low, use a name like rst_n (or substitute n with whatever lowercase character you are using to indicate active-low signals).

Rule – When describing multibit buses, use a consistent ordering of bits. For VHDL, use either (x downto 0) or (0 to x). For Verilog, use [x:0] or [0:x]. Using a consistent ordering helps improve the readability of the code and reduces the chance of accidently swapping order between connected buses.

Guideline – Although the choice is somewhat arbitrary, we recommend using [x:0] for multibit signals in Verilog and (x downto 0) for multibit signals in VHDL. We make this recommendation to establish a standard, and thus achieve some consistency across multiple designs and design teams. See Example 5-1.

Example 5-1   Using [x:0] in port declarations

```
module DW_addinc (
    a,
    b,
    ci,
    sum,
    co
);
input    [(`WIDTH-1):0]    a;
input    [(`WIDTH-1):0]    b;
input                      ci;
output   [(`WIDTH-1):0]    sum;
output                     co;
wire     [(`WIDTH-1):0]    a;
wire     [(`WIDTH-1):0]    b;
wire                       ci;
wire     [(`WIDTH-1):0]    sum;
wire                       co;
endmodule
```

Guideline – When possible, use the same name or similar names for ports and signals that are connected (for example, `a => a;` or `a => a_int;`).

Guideline – When possible, use the signal naming conventions listed in Table 5-1.

Table 5-1   Signal naming conventions

Convention	Use
`*_r`	Output of a register (for example, `count_r`)
`*_a`	Asynchronous signal (for example, `addr_strobe_a`)
`*_pn`	Signal used in the *n*th phase (for example, `enable_p2`)
`*_nxt`	Data before being registered into a register with the same name
`*_z`	Tristate internal signal

Consistent naming makes code easier to read—especially by others—and helps in debug, review, maintenance, and modification.

5.2.2   Naming Conventions for VITAL Support

VITAL is a gate-level modeling standard for VHDL libraries and is described in IEEE Specification 1076.4. This specification places restrictions on the naming conventions (and other characteristics) of the port declarations at the top level of a library element.

Normally, an RTL coding style document need not address gate-level modeling conventions. However, some of these issues can affect developers of hard macros. The deliverables for a hard macro include full-functional/full-timing models, where a timing wrapper is added to the RTL code. If the timing wrapper is in VHDL, then it must be VITAL-compliant.

Background

According to IEEE Specification 1076.4, VITAL libraries can have two levels of compliance with the standard: VITAL_Level0 and VITAL_Level1. VITAL_Level1 is more rigorous and deals with the architecture (functionality and timing) of a library cell. VITAL_Level0 is the interface specification that deals with the ports and generics specifications in the entity section of a VHDL library cell. VITAL_Level0 has strict rules regarding naming conventions and port/generic types. These rules were designed so that simulator vendors can assume certain conventions and deal with SDF back-annotation in a uniform manner.

Rules

Section 4.3.1 of IEEE Specification 1076.4 addresses port naming conventions and includes the following rules. These restrictions apply only to the top-level entity of a hard macro.

Rule (hard macro, top-level ports) – Do not use underscore characters (_) in the entity port declaration for the top-level entity of a hard macro.

The reason for the rule above is that VITAL uses underscores as separators to construct names for SDF back-annotation from the SDF entries.

Rule (hard macro, top-level ports) – A port that is declared in entity port declaration shall not be of mode LINKAGE.

Rule (hard macro, top-level ports) – The type mark in an entity port declaration shall denote a type or subtype that is declared in package `std_logic_1164`. The type mark in the declaration of a scalar port shall denote a subtype of `std_ulogic`. The type mark in the declaration of an array port shall denote the type `std_logic_vector`.

Rule (hard macro, top-level ports) – The port in an entity port declaration cannot be a guarded port. Furthermore, the declaration cannot impose a range constraint on the port, nor can it alter the resolution of the port from that defined in the standard logic package.

5.2.3   State Variable Names

Rule – Use a distinctive suffix for state variable names. Recommended names are `<name>_cs` for the current state, and `<name>_ns` for the next state.

5.2.4   Include Informational Headers in Source Files

Rule – Include a commented, informational header at the top of every source file, including scripts. The header must contain:

- Legal statement: confidentiality, copyright, restrictions on reproduction
- Filename
- Author
- Description of function and list of key features of the module
- Date the file was created
- Modification history including date, name of modifier, and description of the change

Example 5-2 shows a sample HDL source file header.

Example 5-2   Header in an HDL source file

```
--This confidential and proprietary software may be used
--only as authorized by a licensing agreement from
--Synopsys Inc.
--In the event of publication, the following notice is
--applicable:
--
-- (C) COPYRIGHT 1996 SYNOPSYS INC.
-- ALL RIGHTS RESERVED
--
-- The entire notice above must be reproduced on all
-- authorized copies.
--
-- File        : DWpci_core.vhd
-- Author      : Jeff Hackett
-- Date        : 09/17/96
-- Version     : 0.1
-- Abstract    : This file has the entity, architecture
--                and configuration of the PCI 2.1
--                MacroCell core module.
--                The core module has the interface,
--                config, initiator,
--                and target top-level modules.
--
-- Modification History:
-- Date        By    Version   Change Description
--
=============================================================
-- 9/17/96    JDH      0.1     Original
-- 11/13/96   JDH              Last pre-Atria changes
-- 03/04/97   SKC              changes for ism_ad_en_ffd_n
--                             and tsm_data_ffd_n
--
=============================================================
```

The informational header shown in Example 5-2 contains key information for anyone reviewing, debugging, or modifying the file on what the file is (in the context of the design), what it is supposed to do, and so on.

5.2.5   Use Comments

Rule – Use comments appropriately to explain processes, functions, and declarations of types and subtypes. See Example 5-3.

Example 5-3   Comments for a subtype declaration

```
--Create subtype INTEGER_256 for built-in error
--checking of legal values.
subtype INTEGER_256 is type integer range 0 to 255;
```

Guideline – Use comments to explain ports, signals, and variables, or groups of signals or variables.

Comments should be placed logically, near the code that they describe. Comments should be brief, concise, and explanatory. Avoid "comment clutter"; obvious functionality does not need to be commented. The key is to describe the intent behind the section of code. Insert comments before a processes, rather than embedded in it, in order not to interrupt the flow of the code.

5.2.6   Keep Commands on Separate Lines

Rule – Use a separate line for each HDL statement. Although both VHDL and Verilog allow more than one statement per line, the code is more readable and maintainable if each statement or command is on a separate line.

5.2.7   Line Length

Guideline – Keep the line length to 72 characters or less.

Lines that exceed 80 characters are difficult to read in print and on standard terminal width computer screens. The 72 character limit provides a margin that enhances the readability of the code and allows space for line numbers.

For HDL code (VHDL or Verilog), use carriage returns to divide lines that exceed 72 characters and indent the next line to show that it is a continuation of the previous line. See Example 5-4.

Example 5-4   Continuing a line of HDL code

```
hp_req <= (x0_hp_req or t0_hp_req or x1_hp_req or
   t1_hp_req or s0_hp_req or t2_hp_req or s1_hp_req or
   x2_hp_req or x3_hp_req or x4_hp_req or x5_hp_req or
   wd_hp_req and ea and pf_req nor iip2);
```

5.2.8   Indentation

Rule – Use indentation to improve the readability of continued code lines and nested loops. See Example 5-5.

Example 5-5   Indentation in a nested `if` loop

```
if (bit_width(m+1) >= 2) then
  for i in 2 to bit_width(m+1) loop
    spin_j := 0;
    for j in 1 to m loop
      if j > spin_j then
        if (matrix(m)(i-1)(j) /= wht) then
          if (j=m) and (matrix(m)(i)(j) = wht) then
            matrix(m)(i)(j) := j;
          else
            for k in j+1 to m loop
              if (matrix(m)(i-1)(k) /= wht) then
                matrix(m)(i)(k) := j;
                spin_j := k;
                exit;
              end if;
            end loop; -- k
          end if;
        end if;
      end if;
    end loop; -- j
  end loop; -- i
end if;
```

Guideline – Use indentation of 2 spaces. Larger indentation (for example, 8 spaces) restricts line length when there are several levels of nesting.

Guideline – Avoid using tabs. Differences in editors and user setups make the positioning of tabs unpredictable and can corrupt the intended indentation. There are programs available, including language-specific versions of emacs, that will replace tabs with spaces.

5.2.9   Do Not Use HDL Reserved Words

Rule – Do not use VHDL or Verilog reserved words for names of any elements in your RTL source files. Because macro designs must be translatable from VHDL to Verilog and from Verilog to VHDL, it is important not to use VHDL reserved words in Verilog code, and not to use Verilog reserved words in VHDL code.

5.2.10   Port Ordering

Rule – Declare ports in a logical order, and keep this order consistent throughout the design.

Guideline – Declare one port per line, with a comment following it (preferably on the same line).

Guideline – For each interface, declare the ports in the following order:

Inputs:
- Clocks
- Resets
- Enables
- Other control signals
- Data and address lines

Outputs:
- Clocks
- Resets
- Enables
- Other control signals
- Data

Guideline – Use comments to describe groups of ports.

Figure 5-1 and Example 5-6 illustrate the port ordering rule and guidelines.

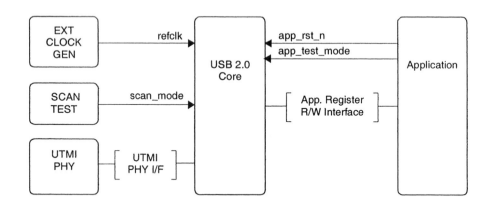

Figure 5-1   USB 2.0 core interfaces

Example 5-6   Port ordering for Figure 5-1

```
module DW_usbd(

// External Clock Generator: Inputs
refclk,                    // Main Reference Clock

Scan Test Interface: Inputs
scan_mode,                 // For scan testing

// UTMI PHY (type 2/3) Interface: Inputs

                           // Enables
phy23_rxvalid,             // Specifies the valid data LSB
phy23_rxvalidh,            // Specifies the valid data on MSB
phy23_txready,             // Valid data will be polled

                           // Other control signals
phy23_linestate,           // Current state of dp, dm lines
phy23_rxerror,             // Error in received data
phy23_rxactive,            // PHY needs to transmit data

                           // Data and address lines
phy23_rx_data,             // 16-bit unidir receive data bus
```

```
// UTMI PHY (type 2/3) Interface: Outputs

                        // Reset
phy23_reset,            // Reset signal to the PHY

                        // Enables
phy23_suspend_n,        // Suspend signal to the PHY
phy23_xcvr_select,      // Select HS or FS transceiver

// Application Interface: Inputs
                        // Resets
app_rst_n,              // Asynchronous reset
app_test_mode,          // Bypassing USB reset

// APP Register R/W Interface: Inputs

                        // Enables
app_16bit,              // APP 16-bit r/w access
app_reg_sel,            // APP Register Interface select

                        // Other control signals
app_rd_n,               // APP register-read command
app_wr_n,               // APP register-write command

                        // Data and address lines
app_addr,               // APP Register address bus
app2usb_data,           // APP Write-data bus

// APP Register R/W Interface: Outputs

                        // Other control signals
usb2app_drdy,           // Data ready indication from
                        // DW_usbd to APP

                        // Data and address lines
usb2app_data,           // APP Read-Data bus
//
// <other APP interface input signals>

// <APP interface output signals>

); // DW_usbd
```

5.2.11   Port Maps and Generic Maps

Rule – Always use explicit mapping for ports and generics, using named association rather than positional association. See Example 5-7.

Example 5-7   Using named association for port mapping

VHDL:

```
U_int_txf_ram : DW_ram_r_w_s_dff
  generic map (
    data_width =>   ram_data_width+ram_be_data_width,
    depth      =>   fifo_depth,
    rst_mode   =>   1
    )
  PORT map (
    clk       =>    refclk,
    rst_n     =>    int_txfifo_ram_reset_n,
    cs_n      =>    logic_zero,
    wr_n      =>    int_txfifo_wr_en_n,
    rd_addr   =>    int_txfifo_rd_addr,
    wr_addr   =>    int_txfifo_wr_addr,
    data_in   =>    int_txfifo_wr_data,
    data_out  =>    txf_ram_data_out
  );
```

Verilog:

```
DW_ram_r_w_s_dff
  #((`ram_data_width+`ram_be_data_width),
    (`fifo_depth),1)
    U_int_txf_ram (
      .clk          (refclk),
      .rst_n        (txfifo_ram_reset_n),
      .cs_n         (1'b0),
      .wr_n         (txfifo_wr_en_n),
      .rd_addr      (txfifo_rd_addr),
      .wr_addr      (txfifo_wr_addr),
      .data_in      (txfifo_wr_data),
      .data_out     (txf_ram_data_out)
    );
```

5.2.12 VHDL Entity, Architecture, and Configuration Sections

Guideline – Place entity, architecture, and configuration sections of your VHDL design in the same file. Putting all the information about a particular design in one file makes the design easier to understand and maintain.

If you include configurations in a source file with entity and architecture declarations, you must comment them out for synthesis. You can do this with the   pragma translate_off and pragma translate_on pseudo-comments in the VHDL source file, as shown in Example 5-8.

Example 5-8   Using pragmas to comment out VHDL configurations for synthesis

```
-- pragma translate_off
configuration cfg_example_struc of example is
  for struc
    use example_gate;
  end for;
end cfg_example_struc;
-- pragma translate_on
```

5.2.13 Use Functions

Guideline – Use functions when possible, instead of repeating the same sections of code. If possible, generalize the function to make it reusable.

For example, if your code frequently converts address data from one format to another, use a function to perform the conversion and call the function whenever you need to. See Example 5-9.

Example 5-9   Creating a reusable function

VHDL:

```
-- This function converts the incoming address to the
-- corresponding relative address.

function convert_address
    (input_address, offset : integer)
  return integer is
begin
  -- ... function bodygoes  here ...
end; -- convert_address
```

Verilog:

```verilog
// This function converts the incoming address to the
// corresponding relative address.

function ['BUS_WIDTH-1:0] convert_address;
   input input_address, offset;
   integer input_address, offset;

begin
   // ... function body goes here ...
end
endfunction // convert_address
```

5.2.14   Use Loops and Arrays

Guideline – Use loops and arrays for improved readability of the source code. For example, describing a shift register, PN-sequence generator, or Johnson counter with a loop construct can greatly reduce the number of lines of source code while still retaining excellent readability. See Example 5-10.

Example 5-10   Using loops to improve readability

```
shift_delay_loop: for i in 1 to (number_taps-1) loop
   delay(i) := delay(i-1);
end loop shift_delay_loop;
```

The array construct also reduces the number of statements necessary to describe the function and improves readability. Example 5-11 is an example of a register bank implemented as a two-dimensional array of flip-flops.

Example 5-11   Register bank using an array

```
type reg_array is array(natural range <>) of
   std_logic_vector(REG_WIDTH-1 downto 0);
signal reg: reg_array(WORD_COUNT-1 downto 0);

begin
   REG_PROC: process(clk)
   begin
      if clk='1' and clk'event then
         if we='1' then
            reg(addr) <= data;
```

```
      end if;
    end if;
  end process REG_PROC;

  data_out <= reg(addr);
```

Guideline – Arrays are significantly faster to simulate than for loops. To improve simulation performance, use vector operations on arrays rather than for loops whenever possible. See Example 5-12.

Example 5-12   Using arrays for faster simulation

Poor coding style:

```
function my_xor( bbit : std_logic;
                 avec : std_logic_vector(x downto y) )
  return std_logic_vector is
variable cvec :
  std_logic_vector(avec'range-1 downto 0);
begin
  for i in avec'range loop        -- bit-level for loop
    cvec(i) := avec(i) xor bbit; -- bit-level xor
  end loop;
  return(cvec);
end;
```

Recommended coding style:

```
function my_xor( bbit : std_logic;
                 avec : std_logic_vector(x downto y) )
  return std_logic_vector is
variable cvec, temp :
  std_logic_vector(avec'range-1 downto 0);
begin
  temp := (others => bbit);
  cvec := avec xor temp;
  return(cvec);
end;
```

5.2.15   Use Meaningful Labels

Rule – Label each process block with a meaningful name. This is very helpful for debug. For example, you can set a breakpoint by referencing the process label.

Guideline – Label each process block *<name>*_PROC.

Rule – Label each instance with a meaningful name.

Guideline – Label each instance U_*<name>*.

In a multi-layered design hierarchy, keep the labels short as well as meaningful. Long process and instance labels can cause excessively long path names in the design hierarchy. See Example 5-13.

Rule – Do not duplicate any signal, variable, or entity names. For example, if you have a signal named incr, do not use incr as a process label—use incr_prc instead. This better relates the name to its use and avoids problems that some tools have with duplicated names.

Example 5-13   Meaningful process label

```
-- Synchronize requests (hold for one clock).
SYNC_PROC : process (req1, req2, rst, clk)

... process body here ...

end process SYNC_PROC;
```

5.3   Coding for Portability

The following guidelines address portability issues. By following these guidelines, you can create code that is technology-independent, compatible with various simulation tools, and easily translatable from VHDL to Verilog (or from Verilog to VHDL).

5.3.1   Use Only IEEE Standard Types (VHDL)

Rule (VHDL only) – Use only IEEE standard types.

You can create additional types and subtypes, but all types and subtypes should be based on IEEE standard types. Example 5-14 shows how to create a subtype (`word_type`) based on the IEEE standard type `std_logic_vector`.

Example 5-14   Creating a subtype from `std_logic_vector`

```
--Create new 16-bit subtype
subtype WORD_TYPE is std_logic_vector (15 downto 0);
```

Guideline (VHDL only) – Use `std_logic` rather than `std_ulogic`. Likewise, use `std_logic_vector` rather than `std_ulogic_vector`. The `std_logic` and `std_logic_vector` types provide the resolution functions required for tristate buses. The `std_ulogic` and `std_ulogic_vector` types do not provide resolution functions.

Note – Standardizing on either `std_logic` or `std_ulogic` is more important than which of the two you select. There are advantages and disadvantages to each. Most designers today use `std_logic`, which is somewhat better supported by EDA tools. In most applications, the availability of resolution functions is not required. Internal tristate buses present serious design challenges and should be used only when absolutely necessary. However, at the system level and in those extreme cases where internal tristate buses are required, the resolution functions are essential.

Guideline (VHDL only) – Be conservative in the number of subtypes you create. Using too many subtypes makes the code difficult to understand.

Guideline (VHDL only) – Do not use the types `bit` or `bit_vector`. Many simulators do not provide built-in arithmetic functions for these types. Example 5-15 shows how to use built-in arithmetic packages for `std_logic_vector`.

Example 5-15   Using built-in arithmetic functions for `std_logic_vector`

```
use ieee.std_logic_arith.all;
signal a,b,c,d:std_logic_vector(y downto x);
    c <= a + b;
```

5.3.2   Do Not Use Hard-Coded Numeric Values

Guideline – Do not use hard-coded numeric values in your design. As an exception, you can use the values 0 and 1 (but not in combination, as in 1001). Example 5-16 shows Verilog code that uses a hard-coded numerical value (7) in the "poor coding style" example and a constant (MY_BUS_SIZE) in the "recommended coding style" example.

Example 5-16   Using constants instead of hard-coded values

Poor coding style:

```
wire      [7:0] my_in_bus;
reg       [7:0] my_out_bus;
```

Recommended coding style:

```
`define MY_BUS_SIZE 8
wire      [`MY_BUS_SIZE-1:0] my_in_bus;
reg       [`MY_BUS_SIZE-1:0] my_out_bus;
```

Using constants has the following advantages:

- Constants are more intelligible as they associate a design intention with the value.
- Constant values can be changed in one place.
- Compilers can spot typos in constants but not in hard-coded values. For example, if MY_BUS_LENGTH is misspelled MY_BUS_LEGTH, the error will be flagged; however, if the intended value 0x100001 is given as 0x10001, the error will not be caught.

5.3.3   Packages (VHDL)

Guideline (VHDL only) – Put constant and parameter definition packages in one or a small number of files with names such as *DesignName*_constants.vhd or *DesignName*_parameters.vhd.

5.3.4   Constant Definition Files (Verilog)

Guideline (Verilog only) – Keep constant and parameter definitions in one or a small number of files with names such as *DesignName*_constants.v or *DesignName*_parameters.v.

Source files should not make references to constant definition files with the include statement, since this will cause problems if the core is instantiated multiple times in the design, but with different constant values. We recommended that constant definition files be specified only on tool command lines.

5.3.5   Avoid Embedding Synthesis Commands

Although it is possible to embed synthesis commands directly in the source code, this practice is not recommended. Others who synthesize the code may not be aware of the hidden commands, which may cause their synthesis scripts to produce poor results. If the design is reused in a new application, the synthesis goals may be different, such as a higher-speed version.

There are several exceptions to this rule. In particular, the synthesis directives to turn synthesis on and off must be embedded in the code in the appropriate places. These exceptions are noted in various guidelines throughout this chapter.

5.3.6   Use Technology-Independent Libraries

Guideline – For arithmetic components, use the DesignWare Foundation Library.

These DesignWare components are all high-performance designs that are portable across processes. They provide significantly more portability than custom-designed, process-specific designs. Using these components helps you create designs that achieve high performance in all target libraries.

The DesignWare Foundation Library contains improved architectures for the inferable arithmetic components, such as:

- Adders
- Multipliers
- Comparators
- Incrementers and decrementers
- Sum of product

These architectures provide improved timing performance over the equivalent internal synthesis tool architectures.

The DesignWare Foundation Library also provides additional arithmetic components such as:

- Sin, cos
- Modulus, divide
- Square root
- Arithmetic and barrel shifters

For more information about using DesignWare components, see the *DesignWare Foundation Library Databook* and the *DesignWare User Guide*. Both of these are available through the Synopsys Documentation on the Web page at

http://solvnet.synopsys.com/cgiservlet/aban/cgi-bin/ASP/dow/dow.cgi

Guideline – Avoid instantiating gates in the design. Gate-level designs are very hard to read, and thus difficult to maintain and reuse. If technology-specific gates are used, then the design is not portable to other technologies.

Guideline – If you must use technology-specific gates, then isolate these gates in a separate module. This will make it easier to modify these gates as needed for different technologies.

Guideline – If you must instantiate a gate, use the Synopsys generic technology library, GTECH. This library contains the following technology-independent logical components:

- AND, OR, and NOR gates (2, 3, 4, 5, and 8)
- 1-bit adders and half adders
- 2-of-3 majority
- Multiplexers
- Flip-flops
- Latches
- Multiple-level logic gates, such as AND-NOT, AND-OR, AND-OR-INVERT

5.3.7   Coding For Translation

When translating from Verilog to VHDL, or vice versa, keep in mind these translation limitations that further restrict each language's synthesizable subset.

Allowing Verilog Translation to VHDL

Guideline (Verilog only) – Do not use any logical expressions in component port maps. There is no equivalent capability in VHDL.

Guideline (Verilog only) – Use unique state names across different state machines. This restriction is due to language differences for scope and visibility.

Guideline (Verilog only) – Verilog functions may only reference the function name, function arguments, and local register variables. This restriction is due to language differences for scope and visibility.

Guideline (Verilog only) – Verilog task arguments should only reference (1) arguments of the task, and (2) register variables defined in the task. If the task is defined inside as an `always` block, it may also reference signals local to the module. Use this last feature cautiously, since these "side effects" may make code maintenance difficult. This restriction is due to language differences for scope and visibility.

Allowing VHDL Translation to Verilog

Guideline (VHDL only) – Do not use `generate` statements. There is no equivalent construct in Verilog.

Guideline (VHDL only) – Do not use `block` constructs. There is no equivalent construct in Verilog.

Guideline (VHDL only) – Do not use code to modify `constant` declarations. There is no equivalent capability in Verilog.

5.4   Guidelines for Clocks and Resets

The following sections contain guidelines for clock and reset signals. The basic theory behind these guidelines is that a simple clocking structure is easier to understand, analyze, and maintain. It also consistently produces better synthesis results. The preferred clocking structure is a single global clock and positive edge-triggered flops as the only sequential devices, as illustrated in Figure 5-2.

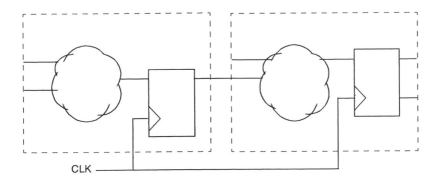

Figure 5-2   Ideal clocking structure

5.4.1   Avoid Mixed Clock Edges

Guideline – Avoid using both positive-edge and negative-edge triggered flip-flops in your design.

Mixed clock edges may be necessary in some designs. In designs with very aggressive timing goals, for example, it may be necessary to capture data on both edges of the clock. However, clocking on both edges creates several problems, and should be used with caution:

• The duty cycle of the clock becomes a critical issue in timing analysis, in addition to the clock frequency itself.

• Most scan-based testing methodologies require separate handling of positive and negative-edge triggered flops.

Figure 5-3 shows an example of a module with both positive-edge and negative-edge triggered flip-flops.

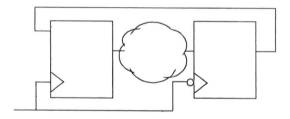

Figure 5-3   Bad example: Mixed clock edges

Rule – If you must use both positive-edge and negative-edge triggered flip-flops in your design, be sure to model the worst-case duty cycle of the clock accurately in synthesis and timing analysis.

The assumption of a perfect clock with 50% duty cycle is optimistic, giving signals half the clock cycle to propagate from one register to the next. In the physical design, the duty cycle will be not be perfect, and the actual time available for signals to propagate can be much smaller.

Rule – If you must use both positive-edge and negative-edge triggered flip-flops in your design, be sure to document the assumed duty cycle in the user documentation.

In most chip designs, the duty cycle is a function of the clock tree that is inserted into the design; this clock tree insertion is usually specific to the process technology. The chip designer using the macro must check that the actual duty cycle will match requirements of the macro, and must know how to change the synthesis/timing analysis scripts for the macro to match the actual conditions.

Guideline – If you must use a large number of both positive-edge and negative-edge triggered flip-flops in your design, it may be useful to separate them into different modules. This makes it easier to identify the negative-edge flops, and thus to put them in different scan chains.

Figure 5-4 shows an example design where the positive-edge triggered flip-flops and negative-edge triggered flip-flops are partitioned into separate blocks.

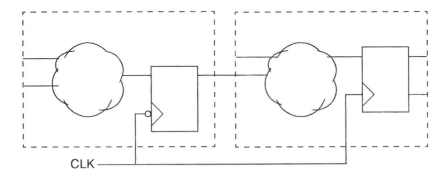

Figure 5-4   Better example: Negative-edge and positive-edge flip-flops are separated

5.4.2   Avoid Clock Buffers

Guideline – Avoid hand instantiating clock buffers in RTL code. Clock buffers are normally inserted after synthesis as part of the physical design. In synthesizable RTL code, clock nets are normally considered ideal nets, with no delays. During place and route, the clock tree insertion tool inserts the appropriate structure for creating as close to an ideal, balanced clock distribution network as possible.

5.4.3   Avoid Gated Clocks

Guideline – Avoid coding gated clocks in your RTL. Clock gating circuits tend to be technology specific and timing dependent. Improper timing of a gated clock can generate a false clock or glitch, causing a flip-flop to clock in the wrong data. Also, the skew of different local clocks can cause hold time violations.

Gated clocks also cause limited testability because the logic clocked by a gated clock cannot be made part of a scan chain. Figure 5-5 on page 104 shows a design where U2 cannot be clocked during scan-in, test, or scan-out, and cannot be made part of the scan chain.

Gated clocks are required for many low-powered designs, but they should not be coded in the RTL for a macro. See "Gated Clocks and Low-Power Designs" on page 105 for the preferred way of dealing with gated clocks. If individual flip-flops need to be gated within a design, the clock gating should be inserted by a power synthesis tool, so that the RTL remains technology portable.

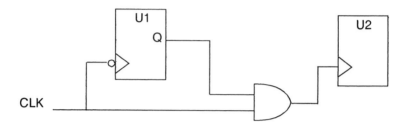

Figure 5-5   Bad example: Limited testability and skew problems because of gated clock

5.4.4   Avoid Internally Generated Clocks

Guideline – Avoid using internally generated clocks in your design.

Internally generated clocks cause limited testability because logic driven by the internally generated clock cannot be made part of a scan chain. Internally generated clocks also make it more difficult to constrain the design for synthesis.

Figure 5-6 shows a design in which U2 cannot be clocked during scan-in, test, or scan-out, and cannot be made part of the scan chain because it is clocked by an internally generated clock. As an alternative, design synchronously or use multiple clocks.

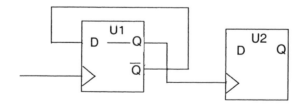

Figure 5-6   Bad example: Internally generated clock

5.4.5   Gated Clocks and Low-Power Designs

Some designs, especially low-power designs, required gated clocks. The following guidelines address this issue.

Guideline – If you must use a gated clock, or an internally generated clock or reset, keep the clock and/or reset generation circuitry as a separate module at the top level of the design. Partition the design so that all the logic in a single module uses a single clock and a single reset. See Figure 5-7.

In particular, clock gating circuitry should never occur within a macro. The clock gating circuit, if required, should appear at the top level of the design hierarchy, as shown in Figure 5-7.

Isolating clock and reset generation logic in a separate module solves a number of problems. It allows submodules 1–3 to use the standard timing analysis and scan insertion techniques. It restricts exceptions to the RTL coding guidelines to a small module than can be carefully reviewed for correct behavior. It also makes it easier for the design team to develop specific test strategies for the clock/reset generation logic.

Guideline – If your design requires a gated clock, model it in RTL using synchronous load registers, as illustrated in Example 5-17 on page 106. This will allow the synthesis tool to insert the actual clock gating logic.

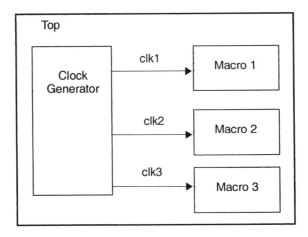

Figure 5-7   Good example: Clock generation circuitry is isolated at the top level

Example 5-17   Use synchronous load instead of combinational gating

Poor coding style:

```
assign clk_p1 = clk and p1_gate;
always @(posedge clk_p1)
  begin : ex5-17a_proc
    . . .
  end // ex5-17a_proc
```

Recommended coding style:

```
always @(posedge clk)
  begin : ex5-17b_proc
    if (p1_gate == 1'b1)
      begin
        . . .
      end
  end // ex5-17b_proc
```

5.4.6   Avoid Internally Generated Resets

Make sure your registers are controlled only by a simple reset signal.

Guideline – Avoid internally generated, conditional resets if possible. Generally, all the registers in the macro should be reset at the same time. This approach makes analysis and design much simpler and easier.

Guideline – If a conditional reset is required, create a separate signal for the reset signal, and isolate the conditional reset logic in a separate module, as shown in Example 5-18 on page 107. This approach results in more readable code and improves synthesis results.

Example 5-18   Isolating conditional reset logic

Poor coding style:

```
always @(posedge clk   or   posedge rst   or   posedge a)
  begin : ex5-18a_proc
    if (rst || a)
      reg_sigs <=  1'b0;
    else begin
      . . .
    end
  end // ex5-18a_proc
```

Recommended coding style:

```
// in a separate reset module
. . .
assign  z_rst = rst || a
  . . .
// in the main module
always @(posedge clk  or posedge z_rst)
  begin : ex5-18b_proc
    if (z_rst)
      reg_sigs <= 1'b0;
    else begin
      . . .
    end
  end // ex5-18b_proc
```

5.4.7   Reset Logic Function

Guideline – The *only* logic function for the reset signal should be a direct clear of all flip-flops. Never use the reset inputs of a flop to implement state machine functionality. Reserving reset pins on flops for reset only makes it easier to generate buffer trees for resets.

5.4.8   Single-Bit Synchronizers

Guideline – Use two flip-flop stages to transfer single bits between clock domains, as shown in Figure 5-8. Label these flip-flops with distinctive names to allow integrators to specify metastability-resistant flip-flops and/or analyze metastability characteristics, it they wish. To prevent transfer of glitches, *do not* connect combinatorial logic from one clock domain to another.

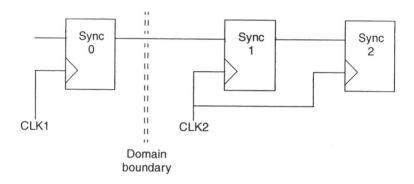

Figure 5-8   Transferring bits between clock domains

5.4.9   Multiple-Bit Synchronizers

Guideline – *Do not* use multiple copies of a single-bit synchronizer (shown in Figure 5-8) to transfer multiple bit fields (such as FIFO address fields) between clock domains. Instead, use a reliable handshake circuit or multibit coding scheme such as Gray code.

5.5   Coding for Synthesis

The following guidelines address synthesis issues. By following these guidelines, you will create code that achieves the best compile times and synthesis results, including:

- Testability
- Performance
- Simplification of static timing analysis
- Gate-level circuit behavior that matches that of the original RTL code

5.5.1   Infer Registers

Guideline – Registers (flip-flops) are the preferred mechanism for sequential logic. To maintain consistency and to ensure correct synthesis, use the templates provided in Example 5-19 (VHDL) and Example 5-20 (Verilog) to infer technology-independent registers. Use the design's reset signal to initialize registered signals, as shown in these examples. In VHDL, do not initialize the signal in the declaration; in Verilog, do not use an `initial` statement to initialize the signal. These mechanisms can cause mismatches between pre-synthesis and post-synthesis simulation.

Example 5-19   VHDL template for sequential processes

```
-- process with asynchronous reset
ex5-19_proc: process(clk, rst_a)
  begin
    if rst_a = '1' then
      . . .
    elseif (clk'event and clk = '1') then
      . . .
    end if;
  end process ex5-19_proc;
```

Example 5-20   Verilog template for sequential processes

```
// process with asynchronous reset
always @(posedge clk or posedge rst_a)
  begin : ex5-20_proc
    if (rst_a == 1'b1)
      begin
        . . .
      end
    else begin
      . . .
    end
  end // ex5-20_proc
```

5.5.2   Avoid Latches

Rule – Avoid using any latches in your design.

As an exception, you can instantiate technology-independent GTECH D latches. However, all latches must be instantiated and you must provide documentation that lists each latch and describes any special timing requirements that result from the latch.

Register files, memories, FIFOs, and other storage elements are examples of situations in which D latches are permitted.

Note – Use a design checking tool to check for latches in your design.

Example 5-21 illustrates a Verilog code fragment that infers a latch because there is no else clause for the if statement. Example 5-22 illustrates a VHDL code fragment that infers a latch because the z output is not assigned for the when others condition.

Example 5-21   Poor coding style: Latch inferred because of missing else
                    condition

```
always @(a or b)
  begin : ex5-21_proc
    if (a == 1'b1)
      q <= b;
    end // ex5-21_proc
```

Example 5-22   Poor coding style: Latch inferred because of missing z output
                    assignment

```
ex5-22_proc: process (c)
begin
  case c is
    when '0' => q <= '1'; z <= '0';
    when others => q <= '0';
  end case;
end process ex5-22_proc;
```

Example 5-23 illustrates a Verilog code fragment that infers latches because of missing s output assignments for the 2'b00 and 2'b01 conditions and a missing 2'b11 condition.

Example 5-23   Poor coding style: Latches inferred because of missing assignments and missing condition

```
always @(d)
  begin : ex5-23_proc
    case (d)
      2'b00: z <= 1'b1;
      2'b01: z <= 1'b0;
      2'b10: z <= 1'b1; s <= 1'b1;
    endcase
  end // ex5-23_proc
```

Guideline – You can avoid inferred latches by using any of the following coding techniques:

- Assign default values at the beginning of a process, as illustrated for VHDL in Example 5-24.
- Assign outputs for all input conditions, as illustrated for Verilog in Example 5-25 on page 112.
- For VHDL, use else (instead of elsif) for the final priority branch, as illustrated in Example 5-26 on page 112.

Example 5-24   Avoiding a latch by assigning default values

```
COMBINATIONAL_PROC : process (state, bus_request)
begin
  -- intitialize outputs to avoid latches
  bus_hold <= '0';
  bus_interrupt <= '0'
  case (state) ...
  . . .
  . . .
end process COMBINATIONAL_PROC;
```

Example 5-25   Avoiding a latch by fully assigning outputs for all input conditions

Poor coding style:

```
always @ (g   or a   or  b)
  begin : ex5-25a_proc
    if (g == 1'b1)
       q <= 0;
    else if (a == 1'b1)
       q <= b;
  end //  ex5-25a_proc
```

Recommended coding style:

```
always @ (g1   or   g2   or   a   or b)
  begin : ex5-25b_proc
    q <=   1'b0 ;
    if (g1 == 1'b1)
       q <= a;
    else if (g2 == 1'b1)
       q <= b;
  end // ex5-25b_proc
```

Example 5-26   Avoiding a latch by using else for the final priority branch
               (VHDL)

Poor coding style:

```
MUX3_PROC: process (decode, A, B)
begin
  if (decode = '0') then
    C <= A;
  elsif (decode = '1') then
    C <= B;
  end if;
end process MUX3_PROC;
```

Recommended coding style:

```
MUX3_PROC: process (decode, A, B)
begin
  if (decode = '1') then
    C <= A;
  else
    C <= B;
  end if;
end process MUX3_PROC;
```

5.5.3   If you must use a latch

In some designs, using a latch is absolutely unavoidable. For instance, in a PCI design, the team found that it was impossible to comply with the PCI specification for reset behavior without having a latch in the design. In order to achieve testability, the team used the approach in Figure 5-9. They used a mux to provide either the normal function or the input from an I/O pad as data to the mux. The mux was selected by the test mode pin used to enable scan.

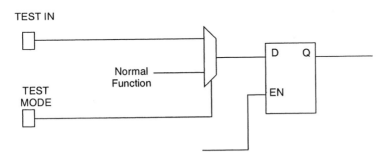

Figure 5-9   Making a latch testable

5.5.4   Avoid Combinational Feedback

Guideline – Avoid combinational feedback; that is, the looping of combinational processes. Combinational feedback causes a number of problems, including making accurate static timing analysis very hard. See Figure 5-10 on page 114.

Bad: Combinational processes are looped

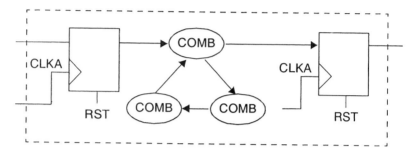

Good: Combinational processes are not looped

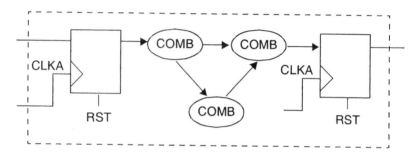

Figure 5-10   Avoiding combinational feedback

5.5.5   Specify Complete Sensitivity Lists

Rule – Include a complete sensitivity list in each of your `process` (VHDL) or `always` (Verilog) blocks.

If you do not use a complete sensitivity list, the behavior of the pre-synthesis design may differ from that of the post-synthesis netlist, as illustrated in Figure 5-11 on page 115.

Synthesis tools as well as design checking tools detect incomplete sensitivity lists and issue a warning when you elaborate the design.

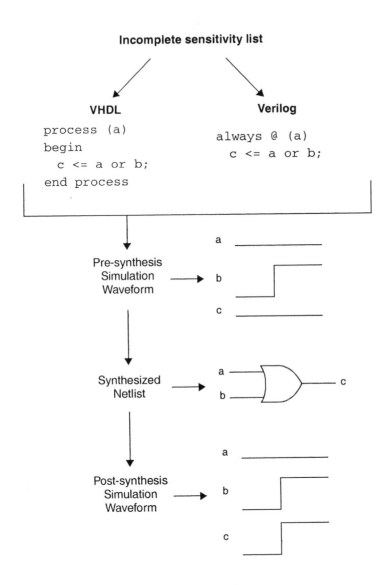

Figure 5-11   Bad example: Simulation mismatch because of incomplete sensitivity list

Combinational Blocks

For combinational blocks (blocks that contain no registers or latches), the sensitivity list must include every signal that is read by the process. In general, this means that every signal that appears on the right side of an assign statement or in a conditional expression. Incomplete sensitivity lists produce incorrect simulation results. See Example 5-27.

Example 5-27   Good coding style: Sensitivity list for combinational process block

VHDL:

```
COMBINATIONAL_PROC : process (a, inc_dec)
begin
  if inc_dec = '0' then
    sum <= a + 1;
  else
    sum <= a - 1;
  end if;
end process COMBINATIONAL_PROC;
```

Verilog:

```
always @(a or inc_dec)
begin : COMBINATIONAL_PROC
  if (inc_dec == 0)
    sum = a + 1;
  else
    sum = a - 1;
end  // COMBINATIONAL_PROC
```

Sequential Blocks

For sequential blocks, the sensitivity list must include the clock signal that is read by the process, as shown in Example 5-28. If the sequential process block also uses an asynchronous reset signal, include the reset signal in the sensitivity list.

Example 5-28   Good coding style: Sensitivity list in a sequential process block

VHDL:

```
SEQUENTIAL_PROC : process (clk)
begin
  if (clk'event and clk = '1') then
    q <= d;
  end if;
end process SEQUENTIAL_PROC;
```

Verilog;

```
always @(posedge clk)
begin : SEQUENTIAL_PROC
  q <= d;
end  // SEQUENTIAL_PROC
```

Sensitivity List and Simulation Performance

Guideline – Make sure your process sensitivity lists contain only necessary signals, as defined in the sections above. Adding unnecessary signals to the sensitivity list slows down simulation.

5.5.6   Blocking and Nonblocking Assignments (Verilog)

In Verilog, there are two types of assignment statements: blocking and nonblocking. Blocking assignments execute in sequential order, nonblocking assignments execute concurrently.

Rule (Verilog only) – When writing synthesizable code, always use nonblocking assignments in `always@(posedge clk)` blocks. Otherwise, the simulation behavior of the RTL and gate-level designs may differ. Specifically, blocking assignments can lead to race conditions and unpredictable behavior in simulations. Also, there can be problems with dependencies if these values are used in other processes.

Example 5-29 shows a Verilog code fragment that uses a blocking assignment where *b* is assigned the value of *a*, then *a* is assigned the value of *b*. The result is the circuit shown in Figure 5-12 on page 118, top, where Register A just loops around and reassigns itself every clock tick. Register B is the same result one time unit later.

Example 5-29   Poor coding style: Verilog blocking assignment

```
always @ (posedge clk)
begin
  b = a;
  a = b;
end
```

Example 5-30 shows a Verilog code fragment that uses a nonblocking assignment. Here, *b* is assigned the value of *a* and *a* is assigned the value of *b* at every clock tick. The result is the circuit shown in Figure 5-12, bottom.

Example 5-30   Recommended coding style: Verilog nonblocking assignment

```
always @(posedge clk)
begin
   b <= a;
   a <= b;
end
```

Bad: Circuit built from blocking assignment

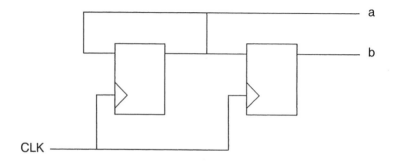

Good: Circuit built from nonblocking assignment

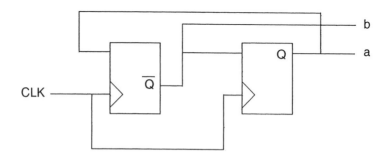

Figure 5-12   Circuit built from blocking and nonblocking assignments

5.5.7 Signal vs. Variable Assignments (VHDL)

In VHDL simulation, signal assignments are scheduled for execution in the next simulation cycle. Variable assignments take effect immediately, and they take place in the order in which they appear in the code. Thus, they present some of the same problems as blocking assignments in Verilog. VHDL variables are not as problematic as Verilog blocking assignments because the interfaces between modules in VHDL are required to be signals, so these interfaces are well-behaved. The order dependencies of variables are thus strictly local, so it is reasonably easy to develop correct code.

Guideline (VHDL only) – When writing synthesizable code, we suggest you use signals instead of variables to ensure that the simulation behavior of the pre-synthesis design matches that of the post-synthesis netlist. If you feel that simulation speed will be significantly improved by using variables, then it is certainly appropriate to do so. Just exercise caution in creating order-dependent behavior in the code.

Note also that in sequential VHDL blocks, variables may or may not result in registers, which complicates debug and maintenance.

Example 5-31   VHDL variable assignment in synthesizable code

Poor coding style:

```
ex5-31a_proc: process (a,b)
variable c : std_logic;
begin
  c := a and b;
end process ex5-31a_proc;
```

Recommended coding style:

```
signal c : std_logic;
ex5-31b_proc : process (a,b)
begin
  c <= a and b;
end process ex5-31b_proc;
```

5.5.8   Case Statements vs. if-then-else Statements

VHDL and fully specified Verilog case statements result in a single-level multi-plexer, while an if-then-else statement infers a priority-encoded, cascaded com-bination of multiplexers. (Note: partially specified Verilog case statements result in latches.)

Figure 5-13 shows the circuit built from the VHDL if-then-else statement in Example 5-32.

Figure 5-14 on page 121 shows the circuit built from the VHDL case statement in Example 5-33.

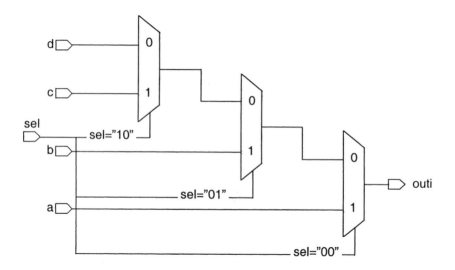

Figure 5-13   Circuit built from if-then-else statement

Example 5-32   Using a VHDL if-then-else statement

```
ex5-32_proc : process (sel,a,b,c,d)
begin
  if (sel = "00") then
    outi <= a;
  elsif (sel = "01") then
    outi <= b;
  elsif (sel = "10") then
    outi <= c;
  else
    outi <= d;
  end if;
```

```
end process ex5-32_proc;
```

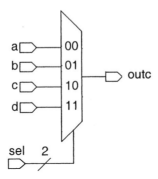

Figure 5-14 Circuit built from the `case` statement

Example 5-33 Using a VHDL `case` statement

```
ex5-33_proc : process (sel,a,b,c,d)
begin
  case sel is
    when "00" => outc <= a;
    when "01" => outc <= b;
    when "10" => outc <= c;
    when others => outc <= d;
  end case;
end process ex5-33_proc;
```

Guideline – The multiplexer is a faster circuit. Therefore, if the priority-encoding structure is not required, we recommend using the `case` statement rather than an `if-then-else` statement. Note that an `if-then-else` statement can be useful if you have a late-arriving signal; this signal can then be connect to the *a* input in Figure 5-14 for the fastest path through the selection function.

In a cycle-based simulator, the `case` statement also simulates faster than the `if-then-else` statement.

A conditional signal assignment may also be used to infer a multiplexer. For large multiplexers, a `case` statement will simulate faster than a conditional assignment on most simulators, and especially on cycle-based simulators. For small muxes, the relative speed of the two constructs varies with different simulators.

Example 5-34 illustrates how to use a conditional assignment to infer a mux.

Example 5-34   Using a conditional assignment to infer a mux

VHDL:

```
z1 <= a when sel_a = '1' else
      b when sel_b = '1' else
      c;

z2 <= d when sel_a = '1' else
      e when sel_b = '1' else
      f;
```

Verilog:

```
assign z1 = (sel_a) ? a : (sel_b) ? b : c;

assign z2 = (sel_a) ? d : (sel_b) ? e : f;
```

5.5.9   Coding Sequential Logic

With current versions of synthesis tools, sequential logic can be written compactly
with a single HDL process. That is, it is no longer necessary to separate all combinational and sequential code.

Guideline – Code sequential logic, including state machines, with one sequential
sequential process. Improve readability by generating complex intermediate variables
outside of the sequential process with assign statements. See Example 5-35.

Example 5-35   State machine with a single process and an intermediate variable

```
// State Machine
// four states,
// intermediate signal generated with assign,
// one registered output,
// one combinatorial output,
// resets to state S0.

reg [1:0] state;
parameter S0 = 2'b00,
          S1 = 2'b01,
          S2 = 2'b10,
          S3 = 2'b11
```

```
// intermediate signal
assign rdy = in_rdy && !wait_1 && (state == s3)

// sequential logic (reset, state logic and
// registered output)
always @(negedge rst_n or posedge clk)
  begin
    if (!rst_n) begin
      state <= S0;
      out1  <= 1'b0;
    end else

      case (state)
        S0: if (input1) begin
              state <= S2;
              out1  <= 1'b1;
            end else begin
              state <= S1;
              out1  <= 1'b0;
            end
        S1: if (rdy) state <= S2;
        S2: state <= S3;
        S3: state <= S0;
      endcase
  end

// combinatorial output
assign out2 = (state == S1) || (state == S2);
```

Guideline – In VHDL, create an enumerated type for the state vector. In Verilog, use `define statements to define the state vector.

Guideline – Keep FSM and non-FSM logic in separate modules if they have different synthesis requirements. For example, a very complex FSM might need to be in a separate module from a complex arithmetic block. See "Partitioning for Synthesis" on page 125 for details.

For more information about coding state machines, read the Optimizing Finite State Machines chapter of the *Design Compiler Reference Manual*. This is available through the Synopsys Documentation on the Web page at

http://solvnet.synopsys.com/cgiservlet/aban/cgi-bin/ASP/dow/dow.cgi

5.5.10   Coding Critical Signals

Guideline – Keep late-arriving signals with critical timing closest to the output of a logic block. For example, use the late-arriving signal early in an `if-else` block.

5.5.11   Avoid Delay Times

Guideline – Do not use any delay constants in RTL code.

The delay values will be incorrect in many environments, may cause simulation behavior to differ from synthesis results, and can interfere with RTL simulator code optimization.

Designers may consider these delays necessary for:

- Mixed RTL and gate-level simulation
- RTL simulation with multiple and/or generated clock signals
- Simulation with externally written code that does not follow coding guidelines

In these cases, use nonblocking assignments, without delays, in your design's sequential code (see Example 5-30 on page 118). If necessary, adjust the clock timing external to your RTL block, or adjust the interface timing from/to your code. This will model what happens in the final chip environment.

5.5.12   Avoid full_case and parallel_case Pragmas

Guideline (Verilog only) – Write Verilog case statements in VHDL style—that is, cover all cases with no overlap. Do not use the `full_case` or `parallel_case` pragmas, because these cause a difference in code interpretation between synthesis and simulation.

5.6   Partitioning for Synthesis

Good synthesis partitioning in your design provides several advantages including:

- Better synthesis results
- Faster compile runtimes
- Ability to use simpler synthesis strategies to meet timing

The following sections illustrate several recommended synthesis partitioning techniques.

5.6.1   Register All Outputs

Guideline – For each subblock of a hierarchical macro design, register all output signals from the subblock.

Registering the output signals from each block simplifies the synthesis process because it makes output drive strengths and input delays predictable. All the inputs of each block arrive with the same relative delay. Output drive strength is equal to the drive strength of the average flip-flop.

Figure 5-15 shows a hierarchical design in which all output signals from each block are registered; that is, there is no combinational logic between the registers and the output ports.

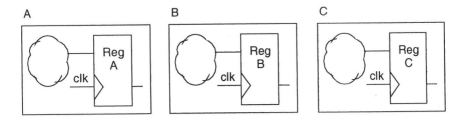

Figure 5-15   Good example: All output signals are registered

5.6.2   Locate Related Combinational Logic in a Single Module

Guideline – Keep related combinational logic together in the same module.

Synthesis tools have more flexibility in optimizing a design when related combinational logic is located in the same module. This is because synthesis tools cannot move logic across hierarchical boundaries during default compile operations.

Figure 5-16 shows an example design where the path from Register A to Register C is split across three modules. Such a design inhibits synthesis tools from efficiently optimizing the combinational logic because it must preserve the hierarchical boundaries in the design.

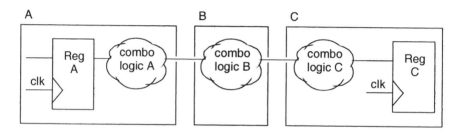

Figure 5-16   Bad example: Combinational logic split between modules

Figure 5-17 shows a similar design in which the related combinational logic is grouped into a single hierarchical block. This design allows synthesis tools to perform combinational logic optimization on the path from Register A to Register C.

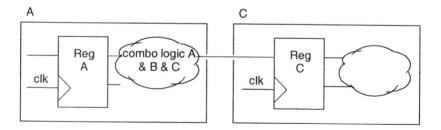

Figure 5-17   Better example: Combinational logic grouped into same module

Figure 5-18 shows an even better design where the combinational logic is grouped into the same module as the destination register. This design provides for improved sequential mapping during optimization because no hierarchical boundaries exist between the sequential logic and the combinational logic that drives it.

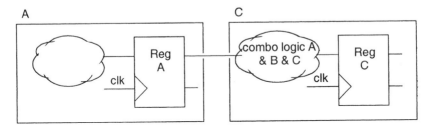

Figure 5-18   Best example: Combinational logic grouped with destination register

Keeping related combinational logic in the same module also eases time budgeting and allows for faster simulation.

5.6.3   Separate Modules That Have Different Design Goals

Figure 5-19 shows a design where the critical path logic is grouped into a separate module from the noncritical path logic. In this design, synthesis tools can perform speed optimization on the critical path logic, while performing area optimization on the noncritical path logic.

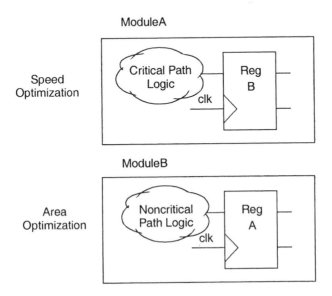

Figure 5-19   Good example: Critical path logic and noncritical path logic
grouped separately

5.6.4   Asynchronous Logic

Guideline – Avoid asynchronous logic.

Asynchronous logic is more difficult to design correctly and verify. Correct timing and functionality may be technology dependent, which limits the portability of the design.

Guideline – If asynchronous logic is required in the design, partition the asynchronous logic in a separate module from the synchronous logic.

Isolating the asynchronous logic in a separate module makes code inspection much easier. Asynchronous logic needs to be reviewed carefully to verify its functionality and timing.

5.6.5   Arithmetic Operators: Merging Resources

A resource is an operator that can be inferred directly from an HDL, as shown in the following code fragment:

```
if ctl = '1' then
   z <= a + b;
else
   z <= c + d;
end if;
```

Normally, two adders are created in this example. If only an area constraint exists, however, synthesis tools are likely to synthesize a single adder and to share it between the two additions. If performance is a consideration, the adders may or may not be merged.

For synthesis tools to consider resource sharing, all relevant resources need to be in the same level of hierarchy; that is, within the same module.

Figure 5-20 is an example of poor partitioning. In this example, resources that can be shared are separated by hierarchical boundaries.

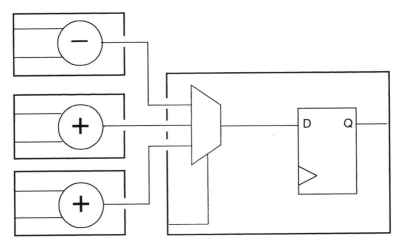

Figure 5-20   Poor partitioning: Resources are separated by hierarchical boundaries

Figure 5-21 is an example of good partitioning because the two adders are in the same module. This partitioning allows synthesis tools full flexibility when choosing whether or not to share the adders.

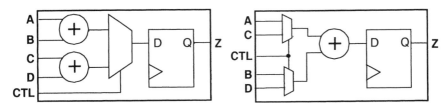

Figure 5-21   Good partitioning: Adders are in the same hierarchy

5.6.6   Partitioning for Synthesis Runtime

In the past, most synthesis guidelines have recommended keeping modules relatively small in order to reduce synthesis runtime. Improvements to synthesis tools, increases in workstation performance, and more experience with large designs has changed this.

The most important considerations in partitioning should be the logic function, design goals, and timing and area requirements. Grouping related functions together is much better than splitting functions artificially, creating complex inter-block timing dependencies. Good timing budgets and appropriate constraints can have a larger impact on synthesis runtime than circuit size does. In one test case, synthesis went from nine hours to 72 hours when the critical range was increased from 0.1 ns to 10 ns.

By grouping logic according to design goals, the synthesis strategy can be focused, reducing synthesis runtime. For example, if the goal for a particular block is to minimize area, and timing is not critical, then the synthesis scripts can be focused on area only, greatly reducing runtime.

Overconstraining a design is one of the biggest causes of excessive runtime. A key technique for reducing runtimes is to develop accurate timing budgets early in the design phase and design the macro to meet these budgets. Then, develop the appropriate constraints to synthesize to this budget. Finally, by developing a good understanding of the synthesis tool commands that implement these constraints, you can achieve an optimal combination of high Quality of Results (QoR) and low runtime.

5.6.7   Avoid Timing Exceptions

A timing exception is a path that does not follow the standard objective of requiring that the data traverse the path in one clock cycle. Examples of timing exceptions are:

- Multicycle paths: Design tools including synthesis, static timing analysis, and place and route tools must allow the data to take *more than one clock cycle* to traverse the path.

- False paths: design tools should *ignore* the amount of time required to traverse the path.

Timing exceptions are problematic because:

- They are difficult to analyze correctly and lend themselves to human error.

- They must be marked as exceptions to all of the design tools, each of which may have its own format and limitations for specifying exceptions.

- The definition of the path at the module level has eventually to be translated to the chip level. The start and end points of the path *may not exist or may not be valid for a particular design* when your module is embedded in a chip, requiring the end user to redefine your design exceptions.

- Most design tools run much slower when many timing exceptions are present.

Guideline – Avoid multicycle paths and other timing exceptions in your design.

Guideline – If you must use timing exceptions in your design, use start and end points which are guaranteed to exist and be valid at the chip level.

Isolating point-to-point exceptions (for example, multicycle paths) within a module improves compile runtime and synthesis results. Also, timing analysis and physical synthesis tools have limited support or poor performance when timing exceptions cross hierarchical boundaries.

Figure 5-22 shows an acceptable multicycle path definition whose start and end points are registers within the same module. These registers are readily identifiable, and will be valid start and end points at the chip level.

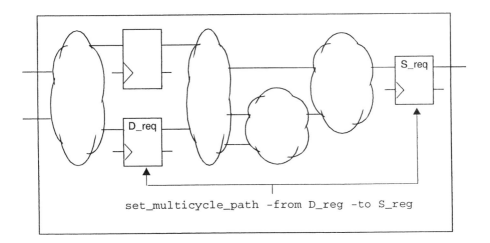

Figure 5-22 Good example: Isolating a point-to-point exception to a single module

Figure 5-23 shows how to handle false paths relative to chip I/Os.

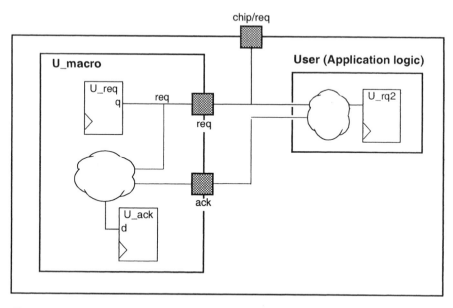

If necessary, use: `set_false_path -from U_req/Q -to U_ack/D`
If necessary, use: `set_false_path -from U_req/Q -to chip/req`
Do not use: `set_false_path -from U_req/Q -to U_macro/req`

Figure 5-23   Handling false paths with pin I/Os

Guideline – Avoid false paths in your design.

False paths are paths that static timing analysis identifies as failing timing, but that the designer knows are not actually failing.

False paths are a problem because they require the designer to ignore a warning message from the timing analysis tool. If there are many false paths in a design, it is easy for the designer accidently to ignore valid warning message about actual failing paths.

If it is necessary to use false paths, follow the multicycle path guidelines in this section.

5.6.8 Eliminate Glue Logic at the Top Level

Guideline – Do not instantiate gate-level logic at the top level of the macro hierarchy.

A design hierarchy should contain gates only at leaf levels of the hierarchy tree. For example, Figure 5-24 shows a design where a NAND gate exists at the top level, between two lower-level design blocks. Optimization is limited because synthesis tools cannot merge the NAND with the combinational logic inside Block C.

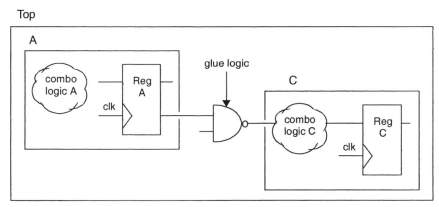

Figure 5-24   Bad example: Glue logic present at top level

Figure 5-25 shows a similar design where the NAND gate is included as part of the combinational logic in Block C. This approach eliminates the extra effort needed to compile small amounts of glue logic and provides for simpler synthesis script development.

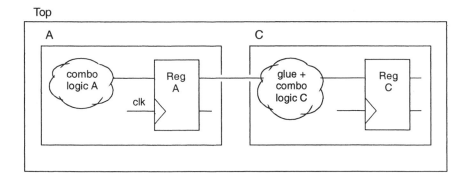

Figure 5-25   Good example: Glue logic grouped into lower-level block

5.6.9   Chip-Level Partitioning

Figure 5-26 shows the partitioning recommendation for the top level of an SoC.

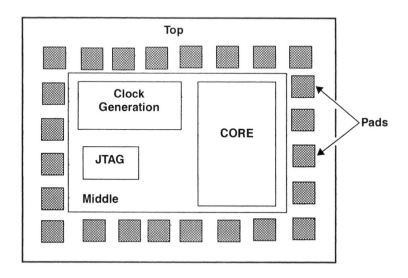

Figure 5-26   Top-level partitioning for an SoC

Guideline – Make sure that the top level of the design contains only the I/O pad ring and clock generator. The next level of hierarchy contains IEEE 1149.1 boundary scan (JTAG) modules and the core logic. The clock generation circuitry is isolated from the rest of the design as it is normally hand crafted and carefully simulated.

5.7   Designing with Memories

Memories present special problems for design reuse, since memory design tends to be foundry specific. Macros must be designed to deal with a variety of memory interfaces. This section outlines some guidelines for dealing with these issues, in particular, designing with synchronous and asynchronous memories.

Synchronous memories present the ideal case, and their interfaces are in the general form shown in Figure 5-27.

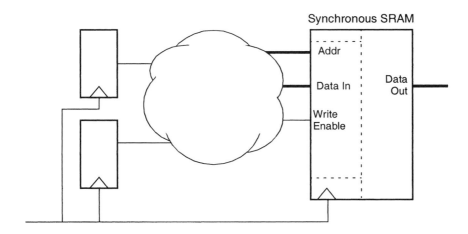

Figure 5-27   Synchronous memory interface

In the past, both synchronous and asychronous RAM cores have been used. Synchronous are the only reliable memories for SoC because asynchronous memories have problems generating write enable pulses---a problem better solved by the RAM designer than the SoC designer.

Guideline – Keep memory interface pins at the top level of a macrocell to allow user choice of memory implementation, interface, and test.

5.8   Code Profiling

In some cases, *code profiling* can assist you in optimizing your code for simulation speed. Some simulators, and several third-party tools, provide the capability of tracking how often each line of code is executed during a given simulation run.

Profiling is a valuable technique that can reveal bottleneck areas in the model. However, you must keep in mind that the profiler looks only at the frequency with which a line is executed, not at how expensive that construct is in terms of machine cycles. For example, performing a variable assignment statement differs a great deal from performing a signal assignment.

| CHAPTER 6 | *Macro Synthesis Guidelines* |

This chapter discusses strategies for developing macro synthesis scripts that enable integrators to synthesize the macro and meet timing goals. The topics include:

- Overview of the synthesis problem
- Synthesis strategies for reusable macros
- High-performance synthesis
- RAM and datapath generators
- Coding guidelines for synthesis scripts

6.1 Overview of the Synthesis Problem

There are some special problems associated with the synthesis of parameterizable soft macros:

- The macro and synthesis scripts must allow the integrator to synthesize the macro and meet timing goals in the final chip.
- The macro must meet timing with the integrator's gate array or standard cell library.
- The macro must meet timing in the integrator's specific configuration of the macro.

This chapter presents a set of tools and methodologies for achieving these goals.

The synthesis guidelines in this chapter are based on many of the same fundamental principles guiding the previous chapter. First and foremost, synthesis and timing design must start at the beginning of the macro design cycle.

That is:

- Functional specifications for the macro must describe the timing, area, and power requirements for the design.

- For larger, timing-critical blocks, detailed technical specifications for the macro and its various subblocks must describe the timing requirements and interfaces in detail, including specifications for input and output delays and loading.

- RTL needs to be coded from the outset to meet both the functional and the timing requirements of the design. Coding for functionality first, and then fixing timing problems later, causes significant delays and poor overall performance in many designs.

If these fundamental guidelines are followed, then synthesis is a straightforward task. Each synthesizable unit or module in the design has a timing budget. Once each module meets this timing budget, the macro is sure to meet its overall timing goals. Synthesis problems become localized, so the difficult problems can be solved on small modules, where they are the most tractable.

6.2   Macro Synthesis Strategy

There are two basic types of synthesis strategies, top-down and bottom-up, but they both rely on the same basic approach: the designer needs to develop a set of constraints for the macro early in the design process. Accurate, appropriate constraints are key to successful synthesis and physical design.

A top-down strategy consists of setting top-level constraints, and letting the synthesis tool partition timing between the component blocks that make up the design. Synthesis is performed on the entire macro in one pass. For many designs under 100K gates, a top-down synthesis strategy usually gives very good results.

A bottom-up strategy consists of setting constraints for each of the blocks that make up the macro. Synthesis is performed on the lowest blocks first, and then performed for higher-level blocks, using the results from the previous steps. For some specific designs, especially those with very aggressive performance goals, a bottom-up synthesis strategy may be required. Designs that are larger than 100K gates usually require a bottom-up synthesis strategy.

The top-down strategy has the advantage of being simpler. If the design has been well partitioned and well designed, the synthesis tool's automatic partitioning of the timing budget can product optimal results.

The bottom-up strategy has the advantage of faster run times, and in some cases the difference can be significant. In addition, for designs with non-obvious timing behavior, manual time budgeting may produce better results.

6.2.1   Macro Timing Constraints

Rule – The basic timing constraints or budget for the macro must be developed as part of the specification process, before the design is partitioned into blocks and before coding begins. This timing budget must be reviewed regularly during the design process to ensure that it is still reasonable and consistent.

The macro timing budget must specify:

- Clock definition
- Setup time requirements for all signals going into the macro
- Clock-to-output delay requirements for all synchronous outputs of the macro
- Input and output delays for all combinational paths through the macro
- Loading budget for outputs and driving cell for inputs
- Operating conditions, including temperature and voltage

Note that combinational paths through the macro are discouraged, because they create non-local synthesis problems that can be very difficult to resolve. Combinational paths must be carefully documented and their timing budgets closely examined to make sure the design constraints can be met. The preferred method for specifying these combinational delays is to specify the input arrival times and the required output time with respect to the clock, assuming the clock is present in the block.

6.2.2   Subblock Timing Constraints

Regardless of whether a top-down or bottom-up strategy is used, it is useful (for complex macros) to develop subblock timing budgets. This exercise forces the designer to address the issue of timing closure early in the design process.

Rule – The basic timing constraints or budget must be developed for each subblock in the macro. This budget must be developed at the time that the design is partitioned into subblocks, and before coding begins. The budget must be reviewed regularly during the design process to ensure that it is still reasonable and consistent.

The subblock timing budget must specify:

- Clock definition
- Wire load model
- Setup time requirements for all signals going into the subblock
- Clock-to-output delay requirements for all synchronous outputs of the subblock
- Input and output delays for all combinational paths through the subblock
- Loading budget for outputs and driving cell for inputs
- Operating conditions, including temperature and voltage

A good nominal starting point for the loading and driving specifications is to use a two-input NAND gate as the driving cell and a flip-flop data input pin as the output load.

Combinational paths through subblocks are discouraged, just as they are at the macro level. In our experience, most synthesis problems arise from these combinational paths.

6.2.3   Synthesis in the Design Process

Synthesis starts as the individual designers are developing the subblocks of the macro, and is initially performed with a single technology library. Later, during the productization phase, the entire macro is synthesized to multiple libraries to ensure portability.

The designer should start running synthesis as soon as the RTL passes the most basic simulation tests. Performing synthesis at this stage allows the designer to identify and fix timing problems early. Because fixing the tough timing problems usually means modifying or restructuring the RTL, it is much better to deal with these problems before the code is completely debugged. Early synthesis also allows the designer to identify the incremental timing costs of new functionality as it is added to the code.

The target at this early stage of synthesis should be to get to within about 10-20% of the final timing budget. This should be close enough to ensure that the RTL code is structured correctly. The additional effort to achieve the timing budget completely is not worth the effort until the code is passing all functional tests. This additional effort will most likely consist of modifying the synthesis scripts and refining the timing budgets.

The subblocks should meet all timing budgets, as well as meeting all functional verification requirements, before being integrated into the macro.

6.2.4   Subblock Synthesis Process

Guideline – The subblock synthesis process consists of three phases:

1. Perform a compile on the subblock, using constraints based on the budget.
2. Perform characterize-compile or budgeting on the whole subblock, to refine the timing constraints and resynthesize the subblock.
3. Iterate if required.

The characterize-compile or budgeting strategy in Step 2 is documented in the *Design Compiler Reference Manual*. This is available through the Synopsys Documentation on the Web page at

http://solvnet.synopsys.com/cgiservlet/aban/cgi-bin/ASP/dow/dow.cgi

6.2.5   Macro Synthesis Process

When the subblocks are ready for integration, we are ready to perform macro-level synthesis.

Guideline – The top-down macro synthesis process consists of two phases:

1. Perform a characterize-compile or budgeting on the whole macro, using the top-level macro constraints.
2. If necessary to meet the timing goals, perform an incremental compile.

Guideline – The bottom-up macro synthesis process consists of three phases:

1. Perform a compile on each of the subblocks, using constraints based on the budget.
2. Perform a characterize-compile or budgeting on the whole macro to improve timing and area.
3. If necessary to meet the timing goals, perform an incremental compile.

The characterize-compile or budgeting is needed to develop accurate estimates of the loading effects on the inputs and outputs of each subblock. Initially, the drive strength of the cells driving inputs, and the loading effects of cells driven by the outputs, are estimated and set manually. The set_driving_cell and set_load commands are used for this purpose. The characterize-compile or budgeting step derives actual drive strengths and loading from the rest of the macro. Clearly, this requires an initial synthesis of the entire macro in order to know what cells are driving/loading any specific subblock input/output.

6.2.6   Wire Load Models

Wire load models estimate the loading effect of metal interconnect upon cell delays. For deep submicron designs, this effect dominates delays, so using accurate wire load models is critical.

The details of how a given technology library does wire load prediction varies from library to library, but the basic principles are the same. A statistical wire length is determined based on the physical size of the block. From this statistical wire length and the total input capacitance of the nodes on the net, the synthesis tool can determine the total load on the driving cell.

The most critical factor in getting an accurate statistical wire length is to estimate accurately the size of the block that will be placed and routed as a unit. Typically, a macro will be placed and routed as a single unit, and the individual subblocks that make up the macro will be flattened within the macro. Thus, the appropriate wire load model is determined by the gate count (and thus area) of the entire macro at the top level.

When we synthesize a subblock, we must use the wire load model for the full macro, not just the subblock. If we use just the gate count of the subblock to determine the wire load model, we will get an optimistic model that underestimates wire delays. When we then integrate the subblocks into the macro and use the correct wire load model, we can run into significant timing problems.

6.2.7   Preserve Clock and Reset Networks

Clock networks are typically not synthesized initially; we rely on physical synthesis to insert a balanced clock tree with very low skew. Asynchronous reset networks are also typically treated as special networks, with physical synthesis inserting the appropriate buffers. These networks need to be identified to the synthesis tool.

Guideline – Set ideal_net on clock and asynchronous reset networks. Also, perform a set_drive 0 (infinite drive) on these nets. Include these required commands in the synthesis scripts for the design. See Example 6-1.

Example 6-1   Using set_ideal_net and set_drive 0 on clock and reset networks

```
set_ideal_net [list clk rst]
set_drive 0 [list clk rst]
```

6.2.8   Code Checking Before Synthesis

Several checks should be run before synthesis. These checks can spot potential synthesis problems without having to perform a complete compile.

Design Checking Tools

Design checking tools can quickly check for many different potential problems, including:

- Presence of latches
- Non-synthesizable constructs like "===" or initial
- Whether a case statement was inferred as a mux or a priority encoder
- Whether all bits in a bus were assigned
- Unused macros, parameters, or variables

Synthesis Tools

Once the RTL passes the design checking tool, the elaboration reports from synthesis tools should be examined to check:

- Whether sequential statements were inferred as flip-flops or latches
- Whether synchronous or asynchronous reset was inferred
- Unexpected flip-flops or latches

A clean elaboration of the design is a critical first step in performing synthesis.

6.2.9   Code Checking After Synthesis

After synthesis, the synthesis tool can be used to run a number of checks. The Design Compiler syntax for some of these is as follows:

Loop Checking
Run report_timing -loops to determine whether there are any combinational loops.

Checking for Latches
Run all_registers -level_sensitive to get a report on latches in the design.

Check for Design Rule Violations
Run check_design to check for missing cells, unconnected ports, and inputs tied high or low.

Verify Testability

Run `check_test` to verify that there are scan versions of all flops, and to check for any untestable structures. Soft macros are typically not shipped with scan flops inserted because scan is usually done on a chip-wide basis rather than block-by-block. Thus, it is essential to verify that scan insertion and automatic test pattern generation (ATPG) will be successful.

As part of the productization phase of the macro development process, full ATPG is run.

Verify Synthesis Results

Use an equivalence-checking tool to verify that the RTL and the post-synthesis netlist are functionally equivalent.

6.3   Physical Synthesis

As silicon technology moves to smaller and smaller geometries, wire load models are becoming increasingly inaccurate in predicting post-layout timing. Physical synthesis, by combining synthesis and placement, provides a much more accurate prediction of the performance that the macro will achieve in an actual chip design. For this reason, it is useful to perform physical synthesis on macros before they are released. For timing-critical macros, it is essential to perform physical synthesis, and to provide data to help the integrator in performing physical synthesis while integrating the macro into the final chip.

6.3.1   Classical Synthesis

In the past, synthesis had no automated interaction with layout. Synthesis generated the netlists without any feedback from floorplanning and place-and-route tools, and there was no opportunity to modify synthesis based on findings during layout. Manual iteration between synthesis and placement was slow and painful. If resynthesis was necessary, layout generally had to be redone from scratch. While this lack of interaction between the synthesis and layout stages was manageable for smaller chip sizes and larger geometries, it has become impractical for today's large, deep submicron SoC designs.

The problems produced by this lack of interactivity between synthesis and layout are exacerbated because, as transistors and cells become faster, cell delays decrease and the percentage of delay due to loading factors increases. Information about physical placement becomes more important for synthesis.

6.3.2   Physical Synthesis

Physical synthesis integrates the synthesis and placement phases of design. In today's technologies, placement is the key physical information needed to predict delays. With the large numbers of routing layers available in current technologies, routing delay is very predictable once placement is known.

Physical synthesis requires the designer to provide the physical constraints for the design as well as the timing constraints. These constraints include:

- Physical size of the macro
- Pin locations
- Grouping information
- Net weightings—which nets are most timing critical

With this information, plus the timing constraints for the design, physical synthesis can synthesize the design through placement, including placing clock and reset buffers and inserting scan flops. Performing static timing analysis on the placed version of the macro can give the macro designer an accurate estimate of the performance the design can achieve.

6.3.3   Physical Synthesis Deliverables

For many designs, physical synthesis is performed as part of developing a test chip. But for designs that have aggressive performance goals, it is necessary to provide some physical synthesis deliverables to the integrator. Since the macro designer can not know the detailed application requirements of the integrator, these deliverables come in the form of application notes and examples of how to do physical synthesis on the macro.

These examples should include sample scripts that show how to define the physical size of the macro and the pin locations, as well as the grouping information and net weightings for the macro.

6.4   RAM and Datapath Generators

Memories and datapaths present a special set of problems for design reuse. Historically, memories and high performance datapaths have been designed at the physical level, making them very technology dependent.

6.4.1   Memory Design

On-chip RAM architectures are differentiated by their architecture and their speed-timing-area tradeoffs.

Flop- and latch-based memories are optimal for small memories—up to about 1000 bits. These memories are appropriate for FIFO designs up to 32x32, for example. The advantage of these memories is that they can be coded in RTL, and thus are technology independent and completely reusable.

Register files are specialized designs optimized for memories in the 1-8K bit range. They use a custom memory cell that is smaller than a standard cell latch, but larger than a true memory cell. They have a larger fixed overhead, in decoders and muxes, than a flop-based design, but this overhead is smaller than that of a true memory. Register files are often used for multi-port memories such as register files for processors—hence their name. The bit cell for register files is typically designed specifically for a given technology. Thus, register files are not technology independent, although they are often easier to port to new processes than are true memories.

Larger memories, optimized for densities over 8K bits, are very specialized designs. The bits cells are designed to be as small as possible, since the bit cell array dominates the area of the memory. These bit cells routinely violate the process design rules in order to achieve the desired density, and require special waivers from the normal design rule checking for the process. These memories have rather large fixed overhead in terms of decoders, muxes, and sense amps, but this overhead is still small compared to the area of the bit cell array. They are typically available with one read and one write port, or with two read/write ports. Specialized memories are available with more ports for specific applications. These dense memories are extremely technology-specific, because of the aggressive design of the bit cell. These designs have complex internal timing due to the use of precharge and sense amps. As a result, they are not only technology-specific, but are also challenging to port to new processes.

Most memory providers offer several versions of their large memories, with bit cells optimized either for power or for speed.

The fact that register files and memories are not technology-independent places a significant burden on the developer of reusable designs. In Chapter 5, we described some approaches for dealing with memories in designing reusable macros. Whereas flop-based memories can be included in a macro, designs needing larger memories require interfaces that allow the integrator to connect process-specific memories external to the macro.

6.4.2 RAM Generator Flow

Figure 6-1 shows the typical work flow for the user of a RAM generator. The designer:

- Describes the memory configuration, through either a GUI or a command-line interface. The designer selects the family of memory, typically trading off power, area, and speed.
- Invokes the memory compiler, which produces a simulation model and synthesis model for the memory, as well as physical layout information.
- Performs simulation with models for the rest of the system to verify the functionality of the memory interfaces.
- Performs synthesis with the synthesis model for the RAM and the RTL for the rest of the design. The synthesis model for the RAM is key in determining overall chip timing and allowing optimal synthesis of the modules that interface to the RAM.

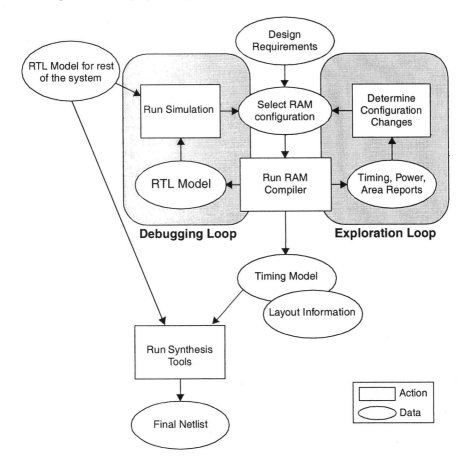

Figure 6-1 RAM generator work flow

6.4.3   Datapath Design

In those datapaths that are dominated by arithmetic functions, the functionality of the design is usually straightforward. The functionality of a 32-bit multiply-accumulate block, for example, is clear and does not help differentiate a design. In order to have a 32-bit MAC that is superior to a competitor's, it is necessary to exploit hardware structure to achieve a faster, smaller, or lower-power design. Historically, this approach has led to tools and methodologies designed to exploit the structural regularity in the datapath, and thus derive a superior physical layout.

Datapath Design Issues

There are three major problems with traditional approaches to datapath design. First, irregular structures like Wallace tree multipliers can outperform regular structures. Second, the datapath designs produced are not portable to new technologies and do not lend themselves to reuse. Third, the great majority of modern applications are poor candidates for the traditional approach, which is best suited to datapaths that are relatively simple (few number of operations) and highly regular (uniform bit widths).

If we look at the history of datapath design, a typical datapath in 1988 was a simple, regular datapath, such as a CPU ALU. Regular structures like muxes and adders dominated; bit slicing was used extensively, and was effective in deriving dense, regular layouts. A 32-bit MAC was a separate chip.

In 2002, graphics, video, and digital signal processing applications are the most common datapath designs. Blocks like IDCTs, FIRs, and FFTs are common datapath elements, and a 32-bit MAC is just a small component in the datapath.

The significant increase in applications for complex datapaths, along with intense pressures to reduce development time, has resulted in a desire to move datapath design to a higher level of design abstraction as well as to leverage design reuse techniques.

Datapath design tools and methodologies are rapidly evolving to meet this need.

Datapath Design Tool Evolution

In the past, designers predominately designed datapaths by handcrafting the design. They captured structural information about the design in schematics and then developed a physical layout of the design. The physical design was laid out for a single bit-slice of the datapath, then replicated. For regular datapaths dominated by muxes and adders, this approach produced dense, regular physical designs.

These handcrafted designs exhibit:

- High performance because the methodology effectively exploited the regular structure of the logic
- Low productivity because of the amount of handcrafting required
- Poor portability because the results were design and technology specific

In the more recent past, designers started using layout-oriented datapath design tools. With these tools, structural descriptions of the design are entered either in schematic form or in HDL, but with severe constraints limiting the subset of the language that can be used. These tools automate much of the handcrafting that was done before, such as developing bit-slice layouts and regular structures. The designs result in:

- High performance for regular structures
- Poor performance for irregular, tree-based structures like Wallace-tree multipliers
- Poor performance for structures with varying bit widths, a common characteristic of graphics designs such as IDCTs, digital filters, or any other design employing techniques like saturation, rounding, or normalization
- Moderate productivity because of the automation of design tasks
- Poor portability because designs were still technology-specific

As a result of recent improvements in datapath synthesis technology, as well as the increasing need for portable designs, designers increasingly are using synthesis for datapath design. Datapath synthesis uses generators to create netlists for datapath components; the specific architecture generated depends on the timing and area goals for the design. The key to good datapath synthesis is to have a rich library of component generators that can produce the optimal netlist based on the width of the datapath, the timing and area constraints of the design, and the function of the operator. In addition, the ability to merge operators in a datapath can be particularly useful. For instance, in generating a sum of products, providing partial products to the final adder can save several levels of logic.

The key realization here is that good datapath design starts with a good netlist, not with a bit-slice physical design. Today's datapaths are dominated by tree structures that have little of the regularity of earlier datapaths. For these structures, physical synthesis tools do at least as good a job as hand design, and often better. The key to performance is to develop the best possible detailed structure (the netlist) and then map it onto the available technology (through physical synthesis).

Of course, even more important than achieving an optimal detailed datapath design is developing an optimal high-level datapath architecture. Datapath synthesis is much faster than other datapath design methods, allowing designers to explore multiple datapath architectures before selecting one for implementation. For instance, when designing an IDCT, the designer can experiment with different saturation algorithms, different numbers of multipliers, and different numbers of pipelines. As a result of this exploration, a superior architecture can be developed. This improved architecture can more than compensate for any loss in performance compared to a handcrafted design.

Because they allow superior, technology-independent designs, datapath synthesis tools provide the opportunity to develop reusable datapath designs without sacrificing performance. This capability is essential for the design of reusable blocks for complex chips in datapath-intensive domains such as video, graphics, and multimedia.

With datapath synthesis tools, designs have:

- High performance: Implementation exploration allows superior designs
- High productivity: Fast synthesis times allow very rapid development of very complex designs
- High portability: Because the source is technology independent and can be parameterized, it is very portable across technologies and from design to design

6.5   Coding Guidelines for Synthesis Scripts

Many of the coding guidelines described in Chapter 5 apply equally well to all scripts, including synthesis scripts.

The following rules and guidelines apply particularly to synthesis scripts:

Rule – All scripts, including synthesis scripts, should begin with a header describing the file, its purpose, its author, and its revision history.

Rule – Comments should be used extensively to describe the synthesis strategy being executed.

Rule – All scripts used to build the design should be under revision control and a bug tracking system, just as the source code is.

Guideline – Keep the line length to 72 characters or less.

Lines that exceed 80 characters are difficult to read in print and on standard terminal width computer screens. The 72 character limit provides a margin that enhances the readability of the code and allows space for line numbers.

For dc_shell commands, use a backslash (\) to continue the statement onto the next line if the command exceeds 72 characters and begin the next line with an indent.

Rule — No hard-coded numbers, data values, or filenames should be buried in the body of the script. Variables should be used in the body of the script and their values set at the top of the script.

Rule — No hard-coded paths should appear in any scripts. Scripts with hard-coded paths are not portable because hard-coded paths prevent the script from being reused in other environments.

Rule — Scripts should be as simple as they can be and still meet their objectives. Synthesis scripts that use only the most common commands are more easily understood and modified.

Rule — Common commands such as those defining the library and search paths should reside in a single setup file, usually the .synopsys_dc.setup file, or in another file that can be included in .synopsys_dc.setup. All other synthesis scripts should perform only those unique tasks for which they were written. Having libraries or search paths defined in multiple files makes modification difficult.

Rule — Synthesis scripts for parameterized soft macros need to be tested as thoroughly as any source code. In particular, all statements and all paths through the script must be tested. Some scripting bugs appear only when the script is used to compile the macro in a particular configuration; these bugs must be uncovered before shipping the script to a customer.

Guideline — Run the syntax checker on synthesis scripts before running the script. The syntax checker can spot many of the scripting errors that can cause synthesis tools to halt or produce useless results.

The following command shows how to use the Design Compiler syntax checker:

```
dc_shell -syntax_check -f ./scripts/my_compile.scr
```

CHAPTER 7

Macro Verification
Guidelines

The goal of macro verification is to ensure that the macro is 100% correct in its functionality and timing. In particular, the behavior of the macro must exactly match the functionality and timing described in the functional specification. This chapter discusses issues in simulating and verifying macros, including the importance of reusable testbenches and test suites, and timing verification. The topics are:

- Overview of macro verification
- Testbench design
- Timing verification

7.1 Overview of Macro Verification

Design verification is consistently one of the most difficult and challenging aspects of design. Parameterized, soft macros being designed for reuse present some particular challenges:

- The verification goal must be for zero defects because the macro may be used in anything from a computer game to a mission-critical aerospace application.
- The goal of zero defects must be achieved for all legal configurations of the macro, and for all legal values of its parameters.
- The integration team must be able to reuse the macro-level testbench components and test suites because the macro must be verified both as a stand-alone design and in the context of the final application.
- Because the macro may be substantially redesigned in the future, the entire set of testbenches and test suites must be reusable by other design teams.

- Since some testbench components may be used in system testing, the testbenches must be compatible with the verification tools used throughout the system testing process.

7.1.1   Verification Plan

Because of the inherent complexity and scope of the functional verification task, it is essential that comprehensive functional verification plans be created, reviewed, and followed by the design team. By defining the verification plan concurrently with the RTL design, the design team can develop the verification environment, including test-benches and verification suites, early in the design cycle. Having a clear definition of the criteria that the macro verification must meet before shipment helps to focus the verification effort and to clarify exactly when the macro is ready to ship.

The specific benefits of developing a verification plan early in the design cycle include:

- The act of creating a functional verification plan forces designers to think through what are typically very time-consuming activities prior to performing them.
- A peer review of the functional verification plan allows a pro-active assessment of the entire scope of the task.
- The team can focus efforts first on those areas in which verification is most needed and will provide the greatest payoff.
- The team can minimize redundant effort.
- The engineers on the team can leverage the cumulative experience and knowledge of the entire team.
- A functional verification plan provides a formal mechanism for correlating project requirements to specific verification tests, which, in turn, allows the completeness (coverage) of the test suite to be assessed.
- Early identification of verification tests allows their development to be tracked and managed more effectively.
- A functional verification plan may serve as documentation of the verification tests and testbench—a critical element for the reuse of these items during regression testing and on subsequent projects. This documentation also reduces the impact of unexpected personnel changes midstream during a project.
- The information contained in the functional verification plan enables a separate verification support team to create a verification environment in parallel with the design capture tasks performed by the primary design team. This can significantly reduce the design cycle time.

The verification environment is the set of testbench components (also known as Veri-fication IP, or VIP) such as bus functional models, bus monitors, memory models, along with the structural interconnect of such components with the design-under-test.

It also includes components for test sequencing, checking, and reporting. Creation of such an environment may involve in-house development of some components and/or integration of off-the-shelf models.

The verification plan should be fully described either in the functional specification for the macro or in a separate verification document. This document will be a living document, changing as issues arise and strategies are refined. The plan should include:

- A description of the test strategy, both at the subblock and the top level.
- A detailed description of the simulation environment, including a block diagram.
- A list of testbench components, such as bus functional models and bus monitors. For each component, there should be a summary of key required features. There should also be an indication of whether the component already exists, can be purchased from a third party, or needs to be developed.
- A list of required verification tools, including simulators and testbench creation tools.
- A list of specific tests, along with the objective and estimated size of each. The size estimate can help in estimating the effort required to develop the test.
- An analysis of the key specifications of the macro, and identification of which tests verify each of these specifications.
- A specification of what functionality of the macro will be verified at the subblock level, and what will be verified at the macro level. Subblock tests must be identified as "throw-away" or as part of the final test environment.
- A specification of the target code coverage for each subblock, and for the top-level macro.
- A description of the regression test environment and regression procedure. The regression tests are those verification tests that are routinely run to verify that the design team has not broken existing functionality while adding new functionality.
- A results verification procedure, specifying what criteria will be used to determine when the verification process has been successfully completed.

7.1.2   Verification Strategy

The verification of a macro consists of three major phases:

- Verification of individual subblocks
- Macro verification
- Prototyping

One definition of a macrocell is: a block that is packaged as a unit for reuse, including its verification environment. The verification environments for the subblocks that make up a macrocell are usually not saved for use in later projects or future versions.

The overall verification strategy is to achieve a very high level of test coverage at the macrocell level, ideally achieving 100% confidence in the functionality of the design for all possible configurations of the macro. This is a daunting task, and to achieve it requires a significant investment in developing a robust verification environment and a complete set of tests. A wide array of tools and techniques are used: bus functional models, monitors for response checking, high-level verification languages, automated regression tests, code coverage tools, and more. This entire environment is then archived with the design, to be reused whenever the team needs to modify the design. A subset of this environment may be shipped with the macro to allow the integrator to verify a specific configuration. Components of the environment, such as BFMs and monitors, may be shipped with the macro to facilitate chip-level verification.

The goal of subblock verification is to achieve a sufficient level of functional coverage of the subblock to allow integration into the macrocell. Integration of multiple subblocks is difficult under any circumstances. If the subblocks have numerous functional problems, then the integration can be extremely difficult and time consuming. How to achieve the required level of subblock verification is very design-dependent, and ultimately up to the designer. For some blocks, it is fairly easy to achieve a very high level of verification using a simple HDL testbench. For other blocks, it is impossible to verify much functionality without essentially reproducing the entire rest of the macrocell, along with its verification environment. For these blocks, it is more appropriate to verify the interfaces and basic functionality with a simple testbench, and then do the bulk of the verification at the macrocell level.

In general, bugs are easier to find and to fix at the subblock level than at the macro level: observability and controllability are typically easier for small pieces of code. Robust subblock-level verification can greatly speed up the overall verification of a macrocell.

At each phase of the verification process, the team needs to decide what kinds of tests to run, and what verification tools to use to run them.

The basic types of verification tests include:

Compliance testing

These tests verify that the design complies with the specification. For an industry-standard design, like a PCI interface or an IEEE 1394 interface, these tests also verify compliance to the published specification. In all cases, compliance to the functional specification for the design is checked as fully as possible.

Corner case testing

These tests try to find the complex scenarios, or corner cases, that are most likely to break the design. They focus on the aspects of the design that are most complex, involve the most interaction between blocks, exercise boundary conditions, or are the least clearly specified.

Random testing

For most designs random tests are an essential complement to compliance and corner case testing. Focused tests like the compliance and corner case tests are limited to the scenarios that the engineers anticipate. Random tests can create scenarios that the engineers do not anticipate, and often uncover the most obscure bugs in a design.

Real code testing

One of the most important parts of verifying any design is running the design with real application code. It is always possible for the hardware design team to misunderstand a specification, and design and test their code to an erroneous specification. Running the real application code is a useful way to uncover these errors.

Regression testing

As tests are developed, they should be added to the regression test suite. This regression test suite can then be run on a regular basis during the verification phase of the project. One of the typical problems in verification is that, in the process of fixing a bug, another bug can be inadvertently introduced. The regression test suite can help verify that the existing baseline of functionality is maintained as new features are added and bugs are fixed. It is particularly important that, whenever a bug is detected, the test case for the bug is added to the regression suite.

Property checking

New tools and methods are emerging for doing property checking on a design. Property checking allows the designer to specify certain properties of the design—for instance, that a particular FIFO should never be read when it is empty. Starting with an initial set of stimulus vectors, the automation tools then determine whether a (legal) sequence of input stimulus can ever violate the property. Property checking is still in its infancy, and tools are just beginning to be available, but this technology is very promising for helping improve the quality of verification.

The verification tools available to the macro design team include:

Simulation

Most of the macro verification is performed by simulating at the RTL level. RTL simulators are very mature, with optimizations to take advantage of well-written, synchronous designs.

Although most simulation should be done at the RTL level, some simulation should be run at the gate level. Typically, this is done late in the design cycle, once the RTL is stable and well-verified. Some initialization problems are masked at the RTL level, since RTL simulation uses a more abstract model for registers, and thus does not propagate Xs as accurately as gate-level simulation. Usually only the reset sequence and a few basic functional tests need to be run at the gate level to verify correct initialization.

Hardware Verification Languages

Hardware Verification Languages (HVLs) dramatically aid the task of creating verification IP and reusable testbenches. These languages provide powerful constructs for generating stimulus and checking response. For example, the designer can check that an event occurred some time within a window of clock cycles. HVLs provide mechanisms for generating random tests and for checking functional test coverage. They also provide mechanisms for communication between testbench objects; this feature can be used to coordinate multiple bus functional models.

Verification IP

There is a significant amount of verification IP commercially available. These bus functional models and monitors can be directly integrated into the testbench, and generate the stimulus and response checking needed for macro-level verification.

Code coverage tools

Code coverage tools provide the ability to assess the quality of the verification suite. They can provide information about what parts of the code have been tested, as well as what states and arcs of a finite state machine have been tested. Code coverage is discussed in more detail in a later section of this chapter.

Hardware modeling

A hardware modeler provides an interface between a physical chip and the software simulator, so that stimulus can be applied to the chip and responses monitored within the simulation environment. Hardware modelers allow the designer to compare the simulation results of the RTL design with those of an actual chip. This verification method is very effective for designs where there is a known-good chip whose functionality is being designed into the macro.

Emulation

Emulation provides very fast run times but long compile times and is significantly more expensive and more difficult to use than simulation. It is an appropriate tool for running real code on a large design, but is not a very useful tool for small macro development.

Prototyping

Building an actual prototype chip using the macro is key to verifying functionality. A prototype allows execution of real code in a real application at real-time speeds. A physical chip is not as easy to debug as a simulation of the design, so prototyping should only occur late in the design phase. Once a problem is detected using a prototype, it is usually best to recreate the problem in the simulation environment, and perform the debug there.

7.2   Inspection as Verification

All of the books on code quality state that the fastest, cheapest, and most effective way to detect and remove bugs is by careful inspection of the design and code. Design reviews and code reviews play a key part in the drive towards zero defects.

Unfortunately, almost all of the research on code quality has been done in the area of software rather than hardware. But the software data we have is compelling and is likely to apply, in some general form, to hardware.

In *Applied Software Measurement* [1], Capers Jones reports that code inspections can be twice as effective as any other method in removing defects. In particular, code inspections are much more effective than test and debug for finding bugs. Jones states:

> Inspections tend to benefit project schedules and effort as well as quality. They are
> extremely efficient at finding interface problems between components and in using
> the human capacity for inductive reasoning to find subtle errors that testing will miss.

In our experience, we have found the same to be true for hardware designs as well. Finding bugs by code inspection is much faster than finding the same bugs by debugging the code during simulation.

There are many styles of design and code review, and a number of authors offer data on the advantages of different styles [1,2]. The following paragraphs describe a typical approach to performing design and code reviews.

A design review is a presentation by the designer (or design team) to the rest of the team. The size of the review team can be quite large. The designer provides the specification document to the reviewers ahead of time, so they can read it and come to the meeting well informed. At the meeting the designer reviews the requirements for the design, and describes in some detail how the design meets these requirements.

Design reviews take place at many points during the design cycle; from the beginning, where the specification is clear but the design is just being defined, up to release of the final design. The level of detail varies at each stage. The purpose of the design review is to review the approach to solving the problem, and to make sure that it is sound. There is no useful way to review the detailed implementation with a large number of people simultaneously.

Code reviews, on the other hand, are reviews of the details of the implementation. They typically involve the designer and a very small number of reviewers, often just a single reviewer. The object of the code review is to do a detailed peer review of the code. The reviewer and the designer go through the code line by line, and the reviewer is expected to fully understand the implementation. Often, teams will insist that reviewers are not managers, to maintain the sense of a supportive, collegial review.

Teams have found that reviews work best when the designer knows that the purpose of the review is to help drive quality, and not for assessment of the designer's performance.

Pressman [2] gives results of some interesting studies assessing the optimal number of reviewers for code walkthroughs.

Code reviews are usually done after a subblock has been designed and verified by the designer, and before it is integrated into the macro.

Design and code reviews may be effective for "non-standard" portions of a design including reset logic, clock generation, and asynchronous interfaces.

Static analysis tools such as design checking tools can also help spot defects before going to simulation.

Thus, there is a whole series of static verification methods that can effectively reduce the number of bugs before even starting dynamic, simulation-based verification. In addition, several software methodology books recommend single stepping through code in a debugger as the first step in dynamic verification. This approach is a combination of dynamic and static verification. By stepping through the code, the designer clearly sees how the code actually behaves in great detail, and can spot bugs as they are executed.

We have very limited experience in using this approach in hardware verification, but encourage readers to try it and see if they find it effective. Single stepping through code clearly works only relatively small blocks; stepping through a million-gate design that requires thousands of cycles to do anything interesting is clearly not a useful exercise. But for subblocks of a macro, this could be an effective verification tool.

7.3   Adversarial Testing

Hardware and software teams have found that having a dedicated team of verification specialists can significantly improve the quality of the final product. Subblock or unit testing is done by the designer, and typically much of the macro verification is done by the design team. However, designers often are focused on proving that the design works correctly.

A separate team of verification experts can take a different view; they can focus on trying to prove that the design is broken. The combination of these two approaches usually gives the best results.

It is also useful to have some members of the team who are verification specialists, and who spend time keeping up with the latest tools and methodologies in verification. In the last few years there has been a proliferation of new point tools targeting verification, from the large EDA companies and from start-ups. Just keeping current on these tools, much less integrating them into the design flow, can be a challenge for the design team.

7.4 Testbench Design

Testbench design differs depending on the function of the macro. For example, the top-level testbench for a microprocessor macro would typically execute test programs, while that of a bus interface macro would typically use bus functional models and bus monitors to apply stimulus and check the results. But the basic principles of good testbench design are common for all macros:

- Stimulus should be generated, and response checked, in terms of transactions.
- The testbench should employ reusable verification components wherever possible, rather than coding from scratch.
- All response checking must be automatic.

Rule – All response checking should be done automatically, rather than having the designer view waveforms and determine whether they are correct.

Guideline – For user-configurable macros, stimulus and response should be automatically customized to the design parameters selected by the user.

7.4.1 Transaction-Based Verification

At some abstract level, all macro testbenches tend to look like Figure 7-1 on page 162.

At this abstract level, we can consider macro interfaces to consist of an input interface and an output interface. The purpose of the testbench is to generate a set of inputs to the input ports and checks the outputs at the output ports. The activity at these ports is not random; in most digital systems, there will be a limited set of *transactions* that occur on a given port. These transactions usually have to do with reading or writing data to some storage element (registers, FIFOs, or memories) in the block.

Figure 7-1   Abstract testbench

Stimulus Generation

When we design the macro, we specify the transaction types that are allowed to occur on a given input port; for example, a register write consists of one specific sequence of data, address, and control pins changing; a register read is another sequence; and perhaps a data burst is a third. Aside from the specified transactions, no other sequence of actions on these pins is legal.

Once we have defined the legal set of transaction types on the input ports, we need to generate sequences of these transactions with the appropriate data/address values for testing the macro. We start by analyzing the functionality of the macro to determine useful sequences that will verify that the macro complies with the specification. Then we search for the corner cases of the design: those unique sequences or combinations of transactions and data values that are most likely to break the design. Finally, we set up random tests to check for conditions we did not think of in the other tests.

After we have developed a complete set of tests, we can measure the coverage of the tests—both functional and code coverage. This gives us a good indication of the completeness of the test suite.

Output Checking

Generating test cases is, of course, just the first part of verification. We must check the responses of the design to verify that it is working correctly. This task consists of two parts: verifying that the output transactions are legal, and verifying that they are correct.

Verifying that the transactions are legal is also called protocol checking. We can write a piece of code, often called a monitor, that will check for illegal transactions—that is, check for protocol violations on the ports of the design. For example, if the read/write line is always supposed to transition one clock cycle before data and be stable until one clock cycle after data transitions, then we can check this automatically.

Verifying that the output transactions are correct is more challenging. Two common techniques are:

- Checking the actual output transactions to expected transactions, based on a separate, reference model for the design. This reference model may be a behavioral or transaction-level model for the design, a physical model such as an existing chip, or simply the model in the mind of the verification engineer. Some techniques for creating this reference model are discussed later in this chapter.

- Setting up testbench components (for example, queues) to collect and check for expected transactions.

7.4.2   Component-Based Verification

Verification components are key to implementing a testbench that generates transactions and automatically checks output transactions. These components provide an essential layer of abstraction between the pins of the device and the transactions specified by the verification engineer.

The testbench can take several forms. An interface macro, such as a USB interface, might have a testbench like the one shown in Figure 7-2 on page 164. Here, the USB has been designed as a peripheral for the AMBA AHB bus, and has two interfaces—the USB port itself, and the AHB interface. The USB can act as a slave on the AHB (when the AHB master is writing/reading registers in the macro) or as a master on the AHB (when the USB is writing/reading data to/from a memory on the AHB bus).

The testbench consists of bus functional models and monitors for the two interfaces. The AHB master BFM creates transactions to read and write registers in the USB macro. The AHB slave BFM acts as an memory on the AHB, the target for data transfers to/from the USB. The AHB monitor checks all transactions on the AHB for protocol violations. It can also give some functional coverage information about transactions on the bus.

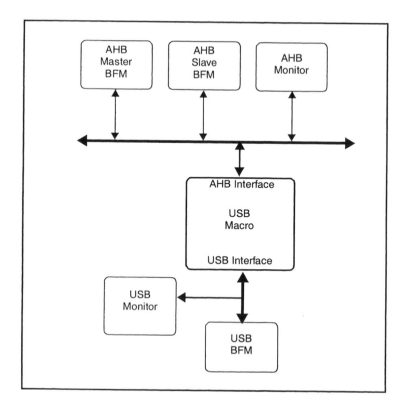

Figure 7-2   Macro development and verification environment

The USB BFM creates transactions on the USB port of the macro. It generates control, bulk, interrupt and isochronous packets, executing the token, data, and handshake phases of each transaction. The USB monitor checks the USB interface for protocol violations and also provides coverage information: has the testsuite generated all kinds of packets, and combinations of packets, and tested multiple endpoints for both IN (read) and OUT (write) transactions.

This kind of component-based verification environment is very powerful. The verification engineer can issue reads and writes to the registers in the USB macrocell without concern for the details of the AHB interface. Similarly, the engineer can generate all the possible transactions on USB interface using very high-level commands rather than constructing packets and phases from scratch.

More importantly, the protocol and coverage monitors in the verification environment provide the verification engineer with important information about what has and hasn't been tested among the possible transactions on the USB interfaces.

One requirement for a complex testbench such as that shown in Figure 7-2 is that actions of the BFMs must be coordinated. HVLs provide message passing and queueing mechanisms so that commands from a central testbench component or program can drive all of the BFMs.

7.4.3   Automated Response Checking

Response Checking with the USB

The bus monitors are very useful for checking the correctness of the basic protocol, but are not adequate to check the full functionality of the USB macro. For this, we need use some form of software or shadow memory to track the data transactions, and make sure they complete successfully.

When the USB BFM is commanded to create a write transaction into the AHB slave memory, the same data is written into a software shadow memory. This second write takes place immediately, while the BFM/USB transaction can take many cycles to execute. Whenever the AHB slave BFM indicates that a write has occurred to its memory, this new value is compared to the value in the equivalent location in shadow memory. In this way the testbench verifies that all the transactions initiated by the USB BFM complete successfully on the AHB bus. In addition, the testbench can verify that the overall transaction completed within some window of time.

This kind of verification approach—coordinating various BFMs, monitors, and shadow memories can be challenging in RTL, but is quite straightforward in an HVL.

Other Approaches

In the previous example, the automated response checking for the testbench was provided by the BFMs, monitors, and shadow memory. This approach is useful for bus interfaces, where data transmission rather than data processing is the design objective. But for other types of macros there are some other techniques that may be useful.

One effective technique is to compare the output responses of the macro to those of a reference design. If the macro is being designed to be compatible with an existing chip (for example, a microcontroller or DSP), then the chip itself can be used as a reference model. A hardware modeler can be used to integrate the physical chip as a model in the simulation environment. Figure 7-3 on page 166 shows a such a configuration.

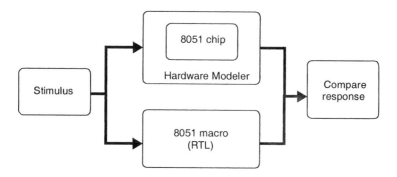

Figure 7-3   Self-checking testbench using a hardware modeler

If a behavioral model for the design was developed as part of the specification process, then this behavioral model can be used as the reference model, especially if it is
cycle-accurate.

One approach often used by microprocessor developers is to develop an Instruction
Set Architecture (ISA) model of the processor, usually in C. This ISA model is
defined to be the reference for the design, and all other representations of the design
must exhibit the same instruction-level behavior. As RTL is developed for the design,
its behavior is constantly being compared to the reference ISA model.

In some sense, the shadow memory in the USB example is a reference design for the
USB macro. The USB is modeled as just passing data through to the memory in the
AHB slave BFM.

7.4.4   Verification Suite Design

Once we have built the testbench, we can develop a set of tests to verify the correct
behavior of the macro. Developing a verification environment and test suite that can
completely verify a macro is even more difficult than specifying or designing the
macro in the first place. Complete verification involves developing a complete
description of the expected behavior of the design, and then proving that the specific
implementation exhibits exactly this behavior.

Compliance Testing

The first step in developing the verification suite is to perform compliance testing; that
is, verifying that the macro implements the functions described in the specification.
This usually involves going through the functional specification essentially line by
line, and developing a test for each required function. The specification (explicitly or
implicitly) defines a set of transactions that can occur at the inputs of the macro, and
what output transactions occur as a result, depending on the internal state of the

macro. Each functional test in the compliance test suite consists of setting the appropriate internal state, applying the appropriate input transactions, and monitoring the output transactions from the macro.

If the specification is an executable specification (for example, a behavioral model), then functional verification involves showing that the behavior of the specification and that of the macro are the same. That is, we need to exercise completely the executable specification, and to show that under the same stimulus, the macro produces the same results. The trick here is to exercise *completely* the executable specification. Running a coverage tool on the executable specification can help to determine the completeness of these tests.

The functional tests in the compliance test suite are necessarily a subset of a complete verification of the macro. The specification does not contain all (and in many cases does not contain any) of the implementation details of the macro. For example, an ISA model for a microprocessor is instruction set accurate, but not cycle-accurate. The cycle-by-cycle behavior of the RTL must be verified in addition to its ability to execute the instruction set correctly.

Compliance testing does establish a baseline of functional correctness for the macro, allowing the verification team to move on to the next stage of testing.

Corner Case Testing

Corner case testing is intended to test the implementation details not covered in the functional testing. In particular, corner case testing focuses on boundary conditions not tested in compliance testing. Designers can often spot corner cases manually. For example, in some microprocessor designs two 32-bit registers can sometimes be used as one 64-bit register. The point where bits roll over from the first 32-bit register to the second is a corner case.

Another typical set of corner cases involve designs with shared resources, such as a bus arbiter, or a block that has multiple agents accessing the same FIFO. For these designs it is useful to create contention for the resources, to ensure that the conflicts are handled correctly.

In general, corner case testing involves testing unusual combinations of data and state: data overflow or underflow, error conditions such as parity generation and checking, and asynchronous behavior. Testing asynchronous behavior can include testing multiple clock domains with extreme clock ratios, to make sure buffering is sufficient to handle the differences in data rates.

While compliance testing gives confidence that the macro works for the most common sets of state and input transactions, corner case testing gives confidence that the macro works correctly for all the conditions the design and verification teams anticipate.

Directed Random Testing and Functional Coverage

Unfortunately, experience has shown that compliance testing and corner case testing are not sufficient to establish complete confidence in a design. There are simply too many scenarios that the design team does not anticipate. For interesting designs, the set of all possible internal states and all possible input transactions is essentially infinite. It is not possible to think of all possible test scenarios, much less test for them.

The best approach to achieve the highest possible confidence in a design is to complement compliance and corner case testing with directed random testing. Over the last few years, design teams have adopted random testing for virtually all designs. Random testing has become an essential verification tool at the macrocell, chip and system level.

Random testing consists of generating random input transactions to the macro. The intent is to have these random transactions occur during random internal states. In this manner, the testing can be evenly distributed over the entire state space of the design, or can be biased to stress specific aspects of the design. This even distribution is key to successful random testing—it is essential to cover as broad a set of conditions as possible.

One very common application of random testing is the case of testing a processor by creating a test program of randomly generated op codes. The behavior of the processor is compared to the behavior of an ISA model for the processor.

Directed random tests are tests that constrain the distribution of tests. For example, when testing a processor, we specify that a certain percentage of instructions should be arithmetic instructions and a different percentage should be load and store, and so on. This kind of directed random testing can be used to assure an optimum distribution of tests.

Functional coverage is another technique for directing random testing. This approach involves defining a set of functional coverage metrics, and directing the random tests to achieve maximum coverage as defined by these metrics. For instance, when testing a processor, we can say that after twenty-five add operations, stop generating add instructions, and constrain testing to the remaining op codes. Similarly, we can say that the test should generate an interrupt during every op code. After a given amount of testing, we can look at the results and see if this metric has been satisfied; if not, we can adjust the distribution of testing to improve coverage.

Generating directed random tests with functional coverage is significantly easier using an HVL rather than an HDL. These verification languages have specific constructs for defining functional coverage and for generating constrained random tests.

7.5   Design of Verification Components

Component-based verification has become the standard method for IP verification as well as for chip and system level verification. At the same time, standard practices have emerged for designing the bus functional models and monitors that are the key components in this approach.

7.5.1   Bus Functional Models

Layers of Abstraction

It is useful to think of bus functional models as operating at up to four levels of abstraction [3]. Lower levels are always in the BFM; higher levels may be implemented in the testbench.

- Layer 0: Signal Layer
- Layer 1: Command Layer
- Layer 2: Traffic Generation Layer
- Layer 3: Test Scenario Generation Layer

Layer 0 provides the basic interface between the model and the device under test. This layer provides the interface between the two languages and their simulators. It also provides the ability to toggle individual pins and to monitor the value of individual pins of the device under test.

Layer 1 provides the first level of commands for the model. These typically are commands such as read and write. In the case of the AHB Master BFM, for example, the command `ahb_write` would cause the model to generate a write transaction on the AHB bus. If necessary, the model would negotiate for access to the bus, and then issue address, data and command signals, and deal with wait states as appropriate. Thus, this layer provides an abstraction of the bus that hides the details of the protocol.

In the case of the USB BFM, the layer 1 commands construct the token, data, and handshake packets that make up the basic USB transactions.

Layer 1 also provides configuration commands for the BFMs. In the case of the AHB slave BFM, for example, this might include commands to configure the model's response to various transaction. One example would be to configure the slave so that reads to address 0x100 in the slave address space produces a one-cycle wait response.

Layer 2 provides commands that use sequences of layer 1 commands to create more complex transactions. In the case of the AHB master BFM, this might include commands to issue back-to-back reads, back-to-back writes, read followed by write, and write followed by read.

In the case of the USB BFM, this might include, for example, a command to generate a bulk data out transaction, consisting of a token packet followed by a data packet, followed by a handshake packet.

Layer 3 provides the highest level of abstraction in the model, allowing the user to generate complex test scenarios. The specific commands for this layer depend greatly on the model and the application. For the AHB master BFM, this might consist of commands to generate random data transfers to various peripherals.

In the case of the USB BFM, this might include commands to generate random data packets over an isochronous connection.

These four layers provide the verification engineer with various levels of abstraction for generating tests. Layer 0 commands might be used to create specific error conditions, such as parity errors. Layer 1 commands might be used to verify the basic functionality of the device under test. Layers 2 might be used to construct complex tests, and Layer 3 to construct directed random testing.

Standard Services

In addition to the four layers of commands, the testbench components provide some standard services to the user. These services include sequence control, messaging, and logging of results.

For pipelined protocols, such as the AHB bus, the model has to issue the address for one command while issuing the data for the previous command. Managing the simultaneous execution of multiple commands is the job of the sequence control service.

The messaging service provides a flexible method for the user to specify what messages should be issued during simulation: for instance, just error, or warning messages, or messages reporting every command that is executed.

The logging service provides a standard mechanism for logging the commands as they are executed. The resulting log file can be useful for debugging the testsuite as well as the design under test.

In addition to these services, it can be useful to have a common method of passing data between the various BFMs and monitors. For instance, as the USB BFM generates an OUT transaction, it may want to inform the AHB monitor, so that it can verify that the transaction occurs on the AHB within a specified time window.

These common resources provide an element of uniformity to the various BFMs used to test IP, helping the verification engineers do their jobs more quickly and efficiently.

7.5.2   Monitors

Monitors provide at least three basic functions:

- Protocol checking
- Coverage reporting
- Transaction logging

In protocol checking, the monitor reports any violations of the protocol for the bus or interface it is monitoring. For example, the AHB specification requires that an error response be held for two cycles. If the AHB monitor detects a one-cycle error response, then it flags it as a protocol violation.

In coverage reporting, the monitor tracks the kinds of transactions that have occurred and generates a report at the end of simulation. For instance, the AHB monitor might generate a report at the end of simulation that gives a count of how many times each transaction type occurred. If this report indicated, for example, that no write was immediately followed by a read, or that no split transaction ever took place, then this might indicate a deficiency in the test suite.

In transaction logging, the monitor provides a trace of activity on the bus or interface it is monitoring. In the case of the AHB monitor, it might generate a log indicating at time 1135 a write transaction to address 0x100 with data 0x1f was initiated, and at time 1145 it was completed. This kind of trace is useful for debugging problems in the design or in the test suite.

7.5.3   Device Models

In addition to bus functional models and monitors, it may be useful to model devices that interface to the device under test. One typical example of this kind of device model is a memory model.

If the macrocell being tested is an SDRAM controller, for example, it is useful to have a full functional model of various SDRAM devices as part of the overall verification testbench. A wide variety of memory models is currently available, using sparse memory modeling techniques to minimize the footprint of the model.

Device models for Physical Layer Interfaces (PHYs) can also be useful. In the case of the USB, for example, the macro may need to interface to several different PHYs. Having accurate functional models for these PHYs can be essential in verifying the USB macro.

7.5.4   Verification Component Usage

The investment in developing or purchasing the verification components needed for
macro verification can be significant. But this kind of component-based verification is
the key to robust verification and to developing testbenches that can be used throughout several generations of macro design.

Fortunately, these verification components are useful at the chip level as well as the
macro level. Once the USB has been integrated into a chip, for example, the USB
BFM and monitor are useful for generating and checking data transactions for chip
level verification. Similarly, the AHB master BFM and monitor can be useful in verifying many different AMBA peripherals and subsystems. At the chip level, the AHB
BFM may eventually be replaced by a processor model, but the monitor will continue
to be useful in chip-level verification.

7.6   Getting to 100%

Unfortunately, delivering a completely bug-free design for any block of even moderate complexity, remains an elusive goal. The techniques described above represent the
state-of-the-art for macro verification, and can deliver a very high level of confidence
in the design. In this section we describe some additional techniques for improving
confidence in the design and managing the risk that some functional bugs remain.

7.6.1   Functional and Code Coverage

Coverage metrics, both functional and code coverage, give the best indication of
whether a design has been fully verified. As mentioned earlier, HVLs allow the engineer to specify measurements of what functionality has been tested. This information
can be used to direct focused or random tests to areas of poor coverage. The test suite
should achieve 100% coverage based on these metrics in order to feel confident that
the entire design has been tested.

Code coverage measures what parts of the RTL code have been tested. This technique
is described in detail below. Code coverage is an important tool for identifying holes
in the test suite, but even 100% code coverage does not mean the design itself has
been fully tested.

7.6.2   Prototyping

Most IP design teams use rapid prototyping to complement simulation, and to compensate for the less-than-100% coverage at the macro verification stage. Achieving
the final small percent of coverage at the macro level is generally extremely costly in

time and still does not detect some of the bugs that will become apparent in prototype operation.

For many macros, it is possible to build a prototype chip and board and thus test the design in the actual target environment. Current FPGA technologies provide the ability to create prototypes very rapidly for all but the largest and most speed-intensive designs. For designs that fit in these technologies and that can be verified at the speeds they provide, these technologies are very useful validation mechanisms.

Building a prototype ASIC is required for macros that must be tested at speeds exceeding those of FPGA technologies.

The major advantage of rapid prototyping is that it gives the team a chance to run actual software on the design. For most designs, simulation is too slow to allow testing of anything beyond functional tests and perhaps the lowest level of software. Prototypes allow running the full application code on the design. For many designs, having application code run is the ultimate test for functional correctness.

7.6.3   Limited Production

Even after robust verification and prototyping, we cannot be sure that there are no remaining bugs in the design. There may be testcases that we did not run, or configurations that we did not prototype. Fundamentally, we have done a robust design but we have not used the macro in a real SoC design. For this reason, we recommend a period of limited production for any new macro. Typically, limited production involves working with just a few (1–4) customers and making sure that they are successful using the macro, before releasing the macro to widespread distribution. We have found this cautious approach very beneficial in reducing the risk of support problems.

7.6.4   Property Checking

All the techniques described above drive our confidence in the design asymptotically towards 100%. Formal methods offer the most promising hope of actually getting there. Property checking, also known as assertion checking, is just starting to be used in complex designs.

Property checking is the technique of describing certain properties of a design, and then using formal methods to prove that these properties hold. These properties may describe characteristics of state machines, such as:

- The state machine can never get into a state that it cannot exit
- Two state machines can never get into deadlock

Properties may also be defined for components of the design, such as:

- No reads are ever performed on an empty FIFO
- No writes are ever performed to a full FIFO

Properties may also be defined for the interfaces of the design, such as:

- Defining the set of legal transactions
- Defining constraints on data values for these transactions

To prove that each property holds, the property checking tool examines the space of all possible, legal transactions on the interfaces of the design to see if there is a sequence that will cause the property to be violated. Clearly this search space can become quite large; dealing with size issues is one of the major challenges facing these tools.

7.6.5   Code Coverage Analysis

Verifying test coverage is essential to the verification strategy; it is a key way to assess, quantitatively, the robustness of the test suite. Several commercial tools are available that provide extensive coverage capabilities.

Types of Coverage Tests

The coverage tools typically provide the following metrics:

- Statement coverage
- Branch coverage
- Condition coverage
- Path coverage
- Toggle coverage
- Triggering coverage

Statement coverage gives a count, for each executable statement, of how many times it was executed.

Branch coverage verifies that each branch in an `if-then-else` or `case` statement was executed.

Condition coverage verifies that all branch sub-conditions have triggered the condition branch. In Example 7-1, condition coverage means checking that the first line was executed with $a = 1$ and that it was executed with $b = 0$, and it gives a count of how many times each condition occurred.

Example 7-1 Condition coverage checks branch condition

```
if (a = '1' or b = '0') then
  c <= '1';
else
  c <= '0';
endif;
```

Path coverage checks which paths are taken between adjacent blocks of conditional code. For example, if there are two successive if-then-else statements, as in Example 7-2, path coverage checks the various combinations of conditions between the pair of statements.

Example 7-2 Path coverage

```
if (a = '1' or b = '0') then
  c <= '1';
else
  c <= '0';
endif;

if (a = '1' and b = '1') then
  d <= '1';
else
  d <= '0';
endif;
```

There are several paths through this pair of if-then-else blocks, depending on the values of a and b. Path coverage counts how many times each possible path was executed.

Triggering coverage checks which signals in a sensitivity list trigger a process.

Trigger coverage counts how many times the process was activated by each signal in the sensitivity list changing value. In Example 7-3 on page 176, trigger coverage counts how many times the process is activated by signal *a* changing value, by signal *b* changing value, and by signal *c* changing value.

Example 7-3   Trigger coverage

```
process (a, b, c)
  . . .
```

Toggle coverage counts how many times a particular signal transitions from '0' to '1', and how many times it transitions from '1' to '0'.

Achieving high code coverage with the macro testbench is a necessary but not sufficient condition for verifying the functionality of the macro. Code coverage does nothing to verify that the original intent of the specification was executed correctly. It also does not verify that the simulation results were ever compared or checked. Code coverage only indicates whether the code was exercised by the verification suite.

On the other hand, if the code coverage tool indicates that a line or path through the code was not executed, then clearly the verification suite is not testing that piece of code.

We recommend targeting 100% statement, branch, and condition coverage. Anything substantially below this number may indicate significant functionality that is not being tested.

Path, toggle, and triggering coverage can be used as a secondary metric. Achieving very high coverage here is valuable, but may not be practical. At times it may be best to examine carefully sections of code that do not have 100% path, toggle, or trigger coverage, to understand why the coverage was low and whether it is possible and appropriate to generate additional tests to increase coverage.

One of the limitations of current code coverage tools is in the area of path coverage. Path coverage is usually limited to adjacent blocks of code. If the design has multiple, interacting state machines, this adjacency limitation means that it is unlikely that the full interactions of the state machines are checked.

Recent Progress in Coverage Tools

Coverage tool providers continue to enhance tool performance on state machines. Tools now can recognize state machines in the RTL, and give the designer useful information about what nodes have been covered, as well as what arcs have been traversed. The tools can also examine pairs of state machines and indicate what pairs of states/arcs have been exercised. This coverage is, of course, limited by the computational power of workstations and the complexity of the state machines, but offers an important step forward.

Tool providers also have provided capabilities for using coverage to minimize regression test suites. One of the historical problem with regression tests is that the tend to grow until runtimes significantly affect the team's ability to verify modifications to the design. Many of the new tests add little incremental coverage over existing tests.

Code coverage can be used to prune the overall test suite, eliminating redundant tests and ordering tests so that the first tests run provide the highest incremental coverage. One tool vendor reports that on a project with Hewlett-Packard, this test pruning approach reduced regression test runtime by 91%. Code coverage tools are still limited in their coverage; see the comments on path coverage above. So, it may be worthwhile running the full regression suite on the final version of the design. But running the pruned suite at a 10x savings in simulation time seems like a very reasonable approach during most of the development cycle of a design.

7.7   Timing Verification

Static timing verification is the most effective method of verifying a macro's timing performance. As part of the overall verification strategy for a macro, the macro should be synthesized using a number of representative library technologies. Static timing analysis should then be performed on the resulting netlists to verify that they meet the macro's timing objectives.

The choice of which libraries to use is a key one. Libraries, even for the same technology, can have significantly different performance characteristics. The libraries should be chosen to reflect the actual range of technologies in which the macro is likely to be implemented.

For macros that have aggressive performance goals, it is necessary to include a trial layout of the macro to verify timing. Pre-layout wire load models are statistical and actual wire delays after layout may vary significantly from these models. Doing an actual layout of the macro can raise the confidence in its abilities to meet timing.

Gate-level simulation is of limited use in timing verification. While leading gate-level simulators have the capacity to handle large designs, gate-level simulation is slow. The limited number of vectors that can be run on a gate-level netlist cannot exercise all of the timing paths in the design, so it is possible that the worst case timing path in the design will never be exercised. For this reason, gate-level timing simulation may deliver optimistic results and is not, by itself, sufficient as a timing verification methodology.

Gate-level simulation is most useful in verifying timing for asynchronous logic. We recommend avoiding asynchronous logic, because it is harder to design correctly, to verify functionality and timing, and to make portable across technologies and applications. However, some designs may require a small amount of asynchronous logic. The amount of gate-level, full-timing simulation should be tailored to the requirements of verifying the timing of this asynchronous logic.

Static timing verification, on the other hand, tends to be pessimistic unless false paths are manually defined and not considered in the analysis. Because this is a manual process, it is subject to human error. Gate-level timing simulation does provide a coarse check for this kind of error.

Guideline – The best overall timing verification methodology is to use static timing analysis as the basis for timing verification. You can then use gate-level simulation as a second-level check for your static timing analysis methodology (for example, to detect mis-identified false paths).

References

1. Jones, Capers. *Applied Software Measurement: Assuring Productivity and Quality.* McGraw Hill Text, 1996

2. Pressman, Roger. *Software Engineering: A Practitioner's Approach.* McGraw Hill Text, 1996.

3. Mohtashemi, Mehdi. *Verification IP Modeling Architecture.* White Paper, Synopsys, Inc. April 2002 Website: http://www.open-vera.com/technical/vip_arch.pdf

CHAPTER 8

Developing Hard Macros

This chapter discusses issues that are specific to the development of hard macros. In particular, it discusses the need for simulation, layout, and timing models, as well as the differing productization requirements and deliverables for hard macros. The topics are:

- Overview
- Hard macro design issues
- Hard macro design process
- Physical design for hard macros
- Model development for hard macros
- Portable hard macros

8.1 Overview

Hard macros are macros that have a physical representation, and are delivered in the form of a GDSII file. As a result, hard macros are more predictable than soft macros in terms of timing, power, and area. However, hard macros do not have the flexibility of soft macros; they provide only a fixed configuration, and are not user-configurable. The porting process of the two forms can also be quite different.

In some sense, however, the distinction between hard and soft macros is artificial. Every macro starts out as soft, for RTL has to be the reference implementation model. Every macro ends up in GDSII, and thus in hard form. The only real distinction between hard and soft macros is at which stage of design the developer hands the macro over to the chip designer. Hard macros are just soft macros that are hardened before they are integrated into the chip design. Of course, this early hardening requires the macro developer to do additional work and provide additional deliverables not required for soft macros.

In this book, we assume the following model for hard macros:

- The macro developer delivers GDSII and a full set of models to the silicon vendor.
- The silicon vendor does the physical design for the chip, including integration of the hard macro. Thus, only the silicon provider actually uses, or has access to, the GDSII for the macro.
- The silicon vendor provides the timing and functional models as well as block outlines and pin locations (for example, LEF) to the chip designer.
- The chip designer uses the timing and functional models for the hard macro while designing the rest of the chip. Typically, these models do not include RTL for the macro. Thus, the models must provide all of the functional and physical information needed to design the chip and verify its timing and functionality.

In a large semiconductor company, the macro developer, silicon vendor, and chip designer may just be different groups within the company. Some large systems houses that do their own physical design may purchase a hard macro directly from a third party provider, thus getting both the GDSII and the models. In this case, the systems house is acting as both the chip designer and as the (fabless) silicon provider. However, the case outlined above is general enough to show the issues and challenges involved in developing high-quality hard macros.

8.1.1   Why and When to Use Hard Macros

Developers typically provide hard versions of macros for any one of several reasons:

- The design is pushing performance to the limit of the silicon process, and thus the physical design must be done by the designer, who knows exactly how to get optimal performance from the design.
- The chip designer wants to reuse a macro without having to perform functional or physical verification.
- The design requires some full-custom design, and so cannot be delivered in soft (that is, synthesizable) form.
- The value of the macro is so great that the macro provider does not want the chip designers to have access to the RTL. That is, hard macros can provide a greater degree of IP protection for the IP provider.
- The macro provider wishes to prevent the possibility of the user modifying the macro.

In the case of processors, all of these conditions are often the case. For this reason, processors are the most common macros to be delivered to the chip designer in hardened form.

There is also a case in which soft macros are used as virtual hard macros. In some very large chips, the design team will use a divide-and-conquer approach to physical design. Each major block, including each soft macro, is placed and routed independently of the other blocks. Chip-level physical design then consists of placing these blocks and wiring them up. In such cases, most of the issues for hard macros described below apply to these independent blocks. In particular, timing and functional models for each of the major blocks can provide more abstract representations of the timing and functionality of the block. These models can provide a faster path to timing convergence and functional verification.

8.1.2 Design Process for Hard vs. Soft Macros

Hard macro development is essentially an extension to soft macro development. The extra steps for hard macros are primarily:

- Generating a physical design
- Developing models for simulation, layout, and timing.

These requirements stem from the fact that hard macros are delivered as a physical database rather than RTL. Integrators require these models to perform system-level verification, chip-level timing, floorplanning, and layout.

Guideline – We recommend that the design process itself be kept identical with the design process for soft macros except for the productization phase.

8.2 Design Issues for Hard Macros

There are several key design issues that are unique to hard macros. These issues affect the design process, and are described in the following sections.

8.2.1 Full-Custom Design

Unlike soft macros, hard macros offer the opportunity to include some full-custom design in a reusable form. However, advances in synthesis, libraries, and timing-driven place and route have largely eliminated the performance advantage for full-custom design. And since full-custom design imposes a significant cost in terms of development schedule, it should only be used in a few, specific circumstances.

Memory is the first and most natural candidate for full-custom implementation. Memory compilers can produce much smaller, faster, lower-power memories than synthesized flop-based memories. We expect all memories except very small FIFOs to be generated from a memory compiler.

For some datapath elements such as barrel shifters, full-custom design techniques can yield slightly smaller designs than synthesizable versions. For the most cost-sensitive or performance-sensitive designs, it may be worth replacing the synthesized version of these blocks with a full-custom version. The designer should be aware, however, that integrating a full-custom component within a macro requires additional work, especially to provide test capability for the component. In the end, the benefit of full-custom components may well be more illusory than real.

There is considerable advantage in minimizing the amount of full-custom logic in a hard macro. Not only does custom logic slow development time, but it also limits the options for porting the design to different processes. Fully synthesizable designs can be ported by repeating synthesis and place and route. Full-custom macros, or full-custom components within macros, need to be ported by physical design tools or by repeating the manual design, place, and route.

Note that even with handcrafted designs, the RTL for the design is the "golden" reference. For all synthesis methods, automated and manual, formal verification should be used to ensure the equivalence between the final physical design and the RTL.

8.2.2   Interface Design

As in most design, good interface design is critical to producing high-quality, easy to integrate hard macros.

Guideline – We strongly recommend registering all of the inputs and outputs of the macro, and clocking them from a single edge of a single clock. In general, the output drivers should be the same for all output pins, and input setup times should be the same for all input pins.

This technique provides a simple and consistent interface for chip designers using the macro, and thus can speed up integration significantly. In addition, consistent timing on ports can simplify synthesis and timing verification scripts for the rest of the chip, reducing the chance of a human error, and speeding up timing convergence.

Registering inputs and outputs can also eliminate some difficult problems in IP security, manufacturing test, and timing modeling, as described later in this chapter.

An additional challenge in interface design is choosing the right output drive strength for output ports. Using too strong a drive strength wastes power and area; for lightly loaded outputs, they can also be slower due to increased intrinsic delay over a smaller buffer. Using too small a drive strength, of course, can result in unacceptable delays when driving long wires to other blocks. Ultimately, this choice is a judgement call, but we recommend erring on the side of too strong a drive strength rather than too weak. Wire delays are only getting greater as technologies shrink.

Registering all outputs helps make designs less sensitive to output drive strengths, especially if the other blocks in the chip register their inputs. In this case, signals have an entire clock cycle to travel from block to block. For large chip designs, where cross-chip delays can be multiple nanoseconds and clock speeds can be hundreds of megahertz, this approach can make the difference between meeting timing and not meeting timing.

8.2.3 Design For Test

Hard macros pose some unique test issues not found in soft macros. With soft macros, the integrator can choose from a variety of test methodologies: full scan, logic BIST, or application of parallel vectors through boundary scan or muxing out to the pins of the chip. The actual test structures are inserted at chip integration, so that the entire chip can have a consistent set of test structures.

Hard macros do not provide this flexibility; test structures must be built into each hard macro. The integrator then must integrate the test strategy of the hard macro with the test strategy for the rest of the chip. It is the task of the hard macro developer to provide an appropriate test structure for the hard macro that will be easy to integrate into a variety of chip-level test structures.

The hard macro developer must choose between full scan, logic BIST, or application of parallel vectors through boundary scan or muxing out to the pins of the chip.

Full scan offers very high test coverage and is easy to use. Tools can be used to insert scan flops and perform automatic test pattern generation. Fault simulation can be used to verify coverage. Thus, scan is the preferred test methodology for hard macros as long as the delay and area penalties are acceptable. For most designs, the slight increase in area and the very slight increase in delay are more than compensated for by the ease of use and robustness of scan.

For some performance-critical designs, such as a microprocessor, a "near full scan" approach is used, where the entire macro is full scan except for the datapath, where the delay would be most costly. For the datapath, only the first and last levels of flops are scanned.

Logic BIST is a variation on the full-scan approach. Where full scan must have its scan chain integrated into the chip's overall scan chain(s), logic BIST uses an LFSR (Linear Feedback Shift Register) to generate the test patterns locally. A signature recognition circuit checks the results of the scan test to verify correct behavior of the circuit.

Logic BIST has the advantage of keeping all pattern generation and checking within the macro. This provides some element of additional security against reverse engineering of the macro. It also reduces the requirements for scan memory in the tester. Logic BIST does require some additional design effort and some increase in die area for the generator and checker, although tools to automate this process are becoming available.

Parallel vectors are used to test only the most timing or area critical designs. A robust set of parallel vectors is extremely time-consuming to develop and verify. If the macro developer selects parallel vector testing for the macro, boundary scan must be included as part of the macro. Boundary scan provides an effective, if slow, way of applying the vectors to the macro without requiring muxing the macro pins out to the chip pins. Requiring the integrator to mux out the pins places an unreasonable burden on the integrator and restricts the overall chip design.

The preferred test methodology is for the hard macro to be fully scan-testable. The full set of test vectors for the scan chain(s) is delivered as part of the overall macro deliverables. The integrator keeps the hard macro scan chain(s) separate from the scan chain(s) of the rest of the chip.

Identifying the scan chains and pins while designing the RTL shortens the physical design cycle and smooths the design flow. For example, scan ports can be pre-allocated for each block when doing place and route. Actual scan insertion needs to be done after placement with physical synthesis. This produces properly ordered scan chains with accurate loading when using muxed scan.

The biggest challenge in integrating a hard macro into a chip-level scan strategy occurs at the interfaces to the hard macro. If there is any combinational logic between the last flops in the surrounding chip and the first flops in the hard macro, then testing this logic is problematic. See "Microprocessor Test" on page 58 for a discussion of how to use shadow registers to solve this problem.

8.2.4   Clock

The hard macro designer has to implement a clock structure in the hard macro without knowing in advance the clocking structure of the chip in which the macro will be used. The designer should provide full clock buffering in the hard macro, and provide a minimal load on the clock input(s) to the macro.

One key problems is that the hard macro will have a clock tree insertion delay; that is, the delay from the clock input pin of the macro, through the clock buffers, before the clock arrives at the internal flops. This delay affects the setup and hold times at the macro's inputs and its clock-to-output delays. The chip designer needs to account for this when integrating the macro into the chip.

In particular, clock insertion delay adds to the macro's clock-to-q delay and input hold time requirements, effectively creating a negative setup time for inputs. The insertion delay can result in hold times as large as several nanoseconds, while a gate delay is only tens of picoseconds. Hold times this large cannot be corrected by adding buffers to inputs.

If the insertion delay is small compared to the clock speed of the chip, then the integrator may be able to accept the increased clock-to-q delay. If so, the integrator can treat the clock input of the hard macro as just another device on the chip's clock tree. The only added work is to add buffers to the hard macro inputs to account for the hold time requirements.

If the insertion delay is significant compared to the clock speed of the chip, then a more aggressive approach is required. In this case, the integrator must deliver an early clock to the macro to compensate for the insertion delay. The integrator can achieve this either during clock tree synthesis, or by using a separate PLL for the hard macro clock.

Rule – The hard macro designer must provide detailed information on the clock insertion delay of the macro.

Most SoC designers today compensate for the insertion delay by providing an early clock to the hard macro. In the future, it may become practical, and necessary, to use a PLL to provide a carefully aligned clock to the macro. At that point, to ease the integration of the macro onto the chip, the hard macro designer should provide a copy of the buffered internal macrocell clock as an output of the macro. This clock is then available to synchronize the macro clock to the system clock via a PLL. Note that in this case, loading on this buffered clock needs to be held to a minimum.

8.2.5   Aspect Ratio

The aspect ratio of the hard macro affects the floorplan and routability of the final chip. Thus, it is an important factor affecting the ease with which the macro can be integrated into the final chip. A large hard macro with an extreme ratio can present significant problems in placing and routing an SoC design. In most cases, an aspect ratio close to 1:1 minimizes the burden on the integrator. Aspect ratios of 1:2 and 1:4 are also commonly used.

Note also that a non-square aspect ratio (for example, a tall, narrow block), means that there will be more routing in vertical direction than in the horizontal. This asymmetric demand on routing resources can lead to problems during place and route. This is another reason why macro designers typically try for a 1:1 aspect ratio.

8.2.6   Porosity

Hard macros can present real challenges to the integrator if they completely block all routing. There are two approaches the hard macro designer can take to mitigate this problem: provide routing channels through the macro, or reserve routing layers above the macro for chip-level routing.

The first approach involves reserving some routing channels through the macro, if it is possible to do so without affecting the macro's performance. This available routing can then be identified in the physical models provided to the integrator.

The second approach is to limit the number of used metal layers to less than the total available in the process. For processes with more than two metal layers available for signal routing, this can be an effective approach to providing routing above the hard macro.

Both of these approaches, however, pose problems. A hard macro or memory is typically characterized for the case where no extra wires are running through or over it. Routing additional wires through or over the block, however, adds capacitance that can slow down adjacent signals. The only way to factor these additional delays into the timing model for the macro or memory is to completely re-characterize the macro or memory in the context of the final chip. In most cases, this re-characterization is not practical. The chip design team has little choice except to hope the additional capacitance does not affect a critical timing path.

For these reasons, designers of leading-edge microprocessors, where each block is treated as a hard macro, leave routing channels between blocks and always route around rather than through blocks.

The guidelines below can facilitate the two approaches described above.

Guideline – The macro deliverables should include a blockage map to identify areas where through-cell routing is possible and will not cause timing problems.

Guideline – The macro deliverables must include a blockage map to identify areas where over-cell routing will not cause timing problems.

Guideline – Leave two layers for over-block routing in complex designs and big hard macros.

Some hard macro designers are using a more extreme approach to this problem. They provide a ground plane as the top layer of the macro. This completely isolates the macro from any chip-level routing on layers above the ground plane, at the cost of consuming an entire routing layer. For small hard macros, this approach may be excessive—it is easier just to route around the macro. But for larger macros, this approach may become more common as geometries shrink and signal integrity issues become more severe.

8.2.7 Pin Placement and Layout

A floorplanning model is one of the deliverables of a hard macro. Among other things, this model describes the pin placement, size, and grid.

Pin placement of the macro can have a significant effect on the floorplan and chip-level routing. Without knowing in detail the target chip design, it is hard to ensure an optimal pin placement. However, common sense suggests that buses and other related signals should be grouped together so that top-level wire lengths can be roughly matched.

The shape and spacing of the physical pins in the hard macro affect how easy it will be for the routing tool to connect the pins to the nets in the chip.

Since the hard macro's orientation can be rotated during placement, the pins should have more than one metal layer type on them. This is because the routing tools prefer to use metal layers in preferred direction to connect to the pins. Although it is possible to use non-preferred direction metal layers to connect to the pins, the router still has to route the pin to the preferred direction which causes congestion and potential via problems.

Rule – Every hard macro pin needs to have at least one horizontal layer and one vertical layer.

Pins on more layers are recommended. For example, if a hard macro utilizes 5 metal layers, the pins could be on M2, M3, M4 and M5.

Pin shapes should all obey the minimum metal area rule. The pin width should be at least 1.5-2x the size of the routing grid in order to cover at least one routing grid. In addition, wherever possible, use the top metal layer rules for all pin layers including spacing.

8.2.8 Power Distribution

Power and ground busing within the macro must be designed to handle the peak current requirements of the macro at maximum frequency. The integrator using the macro must provide sufficient power busing to the macro to limit voltage drop, noise, and simultaneous output switching noise to acceptable levels. The specification of the hard macro must include sufficient information about the requirements of the macro and the electrical characteristics of the power pin contacts on the macro.

The hard macro without power rings is the basic deliverable out of the hardening process. It is optional but recommended that versions of the macro with power rings also be provided as part of the final deliverables. Hard macros that have power rings around them should provide rings in the preferred layer direction, regardless of the orientation of the macro in the final chip design. Thus, two versions of the macro need to be provided: one for the normal placement, and one for 90° rotation placement.

8.2.9   Antenna Checking

Antenna problems occur as a result of the fabrication process. Because metal layers are built up one-by-one during fabrication, a transistor gate can end up with a long piece of metal attached to it which, until other layers are added, is not connected to any path to ground. This piece of metal can act like an antenna, developing a large static charge and damaging the transistor. Antenna rules are used to determine how long an unconnected wire is acceptable.

One of the most painful and time consuming tasks in physical verification is antenna checking and fixing. When routing the final chip, the routing tools do not know or understand the geometries inside the hard macro. Consequently, the routing to and from the clock pins, combined with the internal signal routing can produce antenna violations that the router cannot detect or fix. We recommend inserting diodes of sufficient size on I/O pins to avoid violating the metal-to-diode ratio rule.

Output and Bidirectional Pins

Every output/bidirectional pin of the hard macro should have antenna diodes inserted. The connections from the pin to the diode should be very short and connect down to Metal 1 (M1), as shown in Figure 8-1, top, on page 189.

Use a high-level (top) metal layer, MT (for example, M5 of a six-layer hard macro) for the first wire segment connecting the macro output/bidirectional pin to the net wires inside the hard macro.

Input Pins

For input pins, the hard macro should have not antenna violations when physical verification is run stand-alone.

You should use very short wires from the macro input pin to the receiving cell (or buffer, if the cell is not close), as shown in Figure 8-1, bottom.

Output, bidirectional pins

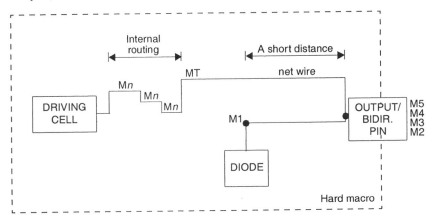

Input pins

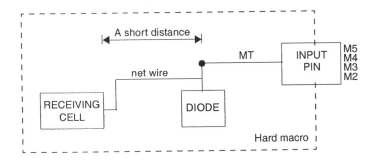

Figure 8-1   Antenna diode insertion for I/O pins

8.3   The Hard Macro Design Process

The hard macro design process is an extension of the process described in Chapter 4. These extensions include additions to the functional specification and the productization of the macro: the physical design of the hard macro, and the generation of models.

We expand the macro specification to include physical design issues. The target library is specified, and timing, area, and power goals are described. The macro specification also addresses the issues described in the previous section: design for test, clock and reset, aspect ratio, porosity, pin placement, and power distribution. The specification describes the basic requirements for each of these. The specification also describes the porting plan: what techniques and tools will be used to port the macro to different processes.

The macro specification also describes the models that will provided as part of the final deliverables. These models include the simulation model(s), timing model(s), and floorplanning model.

8.4   Productization of Hard Macros

Productization of hard macros involves physical design, verification, model development, and documentation. Figure 8-2 on page 192 describes the process of physical design and model development.

8.4.1   Physical Design

The first step in productizing the hard macro is to complete the physical design. Figure 8-2 on page 192 shows the basic loop of floorplanning and incremental synthesis, place and route, and timing extraction.

This process starts with the final floorplan and synthesis of the macro. Then pin placements and initial power routing are done. These steps define the external constraints of the physical design of the macro: the size, aspect ratio, pin locations, and power connections. The power grid created during initial power routing also places internal constraints on placement and routing of the design. Once these are done, the goal of physical design is to complete place and route to meet these constraints, as well as the overall performance goals for the design.

The next step in the process is physical synthesis and scan chain insertion. These steps provide a detailed placement for the design, except for the clock buffers. Based on this placement, clock tree insertion creates the clock tree(s) and places the clock buffers needed to distribute clocks to the design, subject to the skew requirements of the design.

Final power routing is then performed, followed by final detailed routing. This completes the first pass of physical design; now we need to determine whether the design meets its performance goals. To do this, we perform RC extraction and static timing analysis. This process provides the timing performance data for the design. If necessary, we can also perform power analysis to see if the design meets the power goals.

If the physical design does not meet the performance goals, we have two choices. If timing, power, or area is far from meeting specification, we may need to go back to the design phase and iterate as required.

If we are reasonably close to meeting specification, however, we go back to the physical synthesis stage to perform an incremental synthesis, and repeat routing and timing analysis. By retaining as much of the original placement as possible, incremental physical synthesis optimizes our chances of rapidly converging on physical design that meets our performance goals.

Once the design meets the timing goals, we verify other aspects of the design. We perform power analysis to determine whether the IR drop across the design meets requirements, and check that the design does not violate electromigration rules. Also, we run signal integrity analysis to make sure the design makes timing even after crosstalk and other signal integrity issues are taken into account.

Finally, we perform formal verification to assure that the final design is functionally identical to the original RTL.

One key to successful physical design is to have a high quality standard cell library. The library should have been fully characterized to ensure that the timing models are accurate. The library should also have all the views required to complete the design, including power modeling.

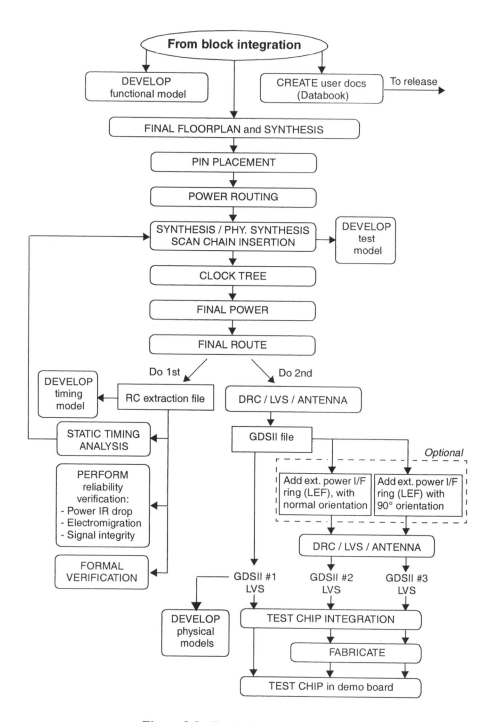

Figure 8-2   Productizing hard macros

8.4.2   Verification

Once we achieve our physical design goals with a place and route, we perform a series of verifications on the physical design:

Gate Verification

We use formal verification to prove that the final gate-level netlist is equivalent to the RTL. For handcrafted blocks, we use a combination of LVS (Layout vs. Schematic), to verify transistor to gate netlist equivalence, and formal verification. We also run full-timing, gate-level simulation to verify any asynchronous parts of the design.

In this book, we strongly recommend fully synchronous design, with no timing exceptions or multicycle paths and as few clock domains as possible. Following these rules makes static timing analysis very straightforward. If these rules are violated, however, it becomes necessary to develop fairly complex scripts to perform timing analysis correctly. These scripts, like all complex scripts, are subject to human error. For these designs, significant gate-level simulation may be necessary as a second check that the design meets timing.

Physical Verification

We use LVS and DRC (Design Rule Checking) tools to verify the correctness of the final physical design. LVS extracts transistors from the physical design and maps these to the original netlist, the assure that the physical design is functionally equivalent to the pre-layout design. DRC checks the physical design to make sure it complies with all the process rules for the technology.

Finally, antenna checks are run, any violations fixed, and DRC is re-run.

The DRC and LVS decks are an important consideration in physical design. These decks provide the physical design rules used in physical verification of the final design. These decks need to be consistent between all the blocks in a chip design; integrating different blocks that use different decks can cause a physical verification nightmare. Typically, the hard macro designer gets these decks from the library provider; it is important that the chip design team uses the same decks for the rest of the chip, and in particular for any other hard macros used in the chip.

8.5   Model Development for Hard Macros

The set of models provided to the integrator fall into the following categories:

- Functional models: used by the integrator to develop and verify the RTL for the rest of the chip
- Timing model: used by the integrator to run static timing analysis on the entire chip
- Power model: used by the integrator to estimate power dissipation and IR drop for the entire chip
- Test model: used by the integrator to develop a complete manufacturing test for the chip
- Physical models: used by the integrator during physical design of the rest of the chip

8.5.1   Functional Models

Hard macros are typically of high value and high complexity; only this kind of design justifies the additional effort to create a hard version. Because of this complexity, the RTL for these designs tends to simulate quite slowly, creating a bottleneck in the design process. Often, hard macros are processors, requiring significant application code to be developed while the chip is being designed. The software developers clearly need a very fast model of the processor to develop this software. On the other hand, the chip designers need a very accurate functional model to be sure that the entire chip will function correctly.

Because of these conflicting needs, it is usually necessary to provide a variety of functional models for a hard macro. These models make various tradeoffs between accuracy and speed to meet the various needs of the hardware and software design teams.

Most of these functional models are created as part of the macro design process. However, the method for packaging and delivering these models tends to be somewhat ad hoc. There are some tools, and some new emerging tools, for automating some aspects of model generation.

Model Security

One of the critical issues in developing a modeling process is determining the level of security required for the models. Source code for functional models can be shipped directly to the customer if security is not a concern. This is often the case for bus functional models, which contain little information about the detailed functionality of the macro. If security is a concern, then some form of protection must be used. Often this security is achieved by providing a compiled version of the model.

One common form of protection is to compile the model and the simulation kernel into a single, stand-alone executable. An RTL wrapper is used to provide a simple timing and functional interface to the RTL for the rest of the chip. The Verilog PLI and VHDL language interfaces provide a reasonably straightforward mechanism for tying this kind of model into the simulator. By delivering object code, the designer ensures a high level of security.

Bus Functional Models

Bus functional models abstract out all the internal behavior of the macro, and only provide the capability of creating transactions on the interfaces of the macro. These models are useful for system simulation when the integrator wants to test the rest of the system, independent of the macro. By abstracting out the internal behavior of the macro, we can develop a very fast model that still accurately models the detailed behavior of the macro at its interfaces.

Bus functional models are also useful in interfacing more accurate models to the simulation of the rest of the chip.

Behavioral and ISA Models

Extensive hardware/software cosimulation is critical to the success of many SoC design projects. In turn, effective hardware/software cosimulation requires very high performance models for large system blocks. Behavioral and ISA models provide this level of performance by abstracting out many of the implementation details of the design.

Most processor design teams develop a high-level model of the processor as they define the processor architecture. This model accurately reflects the instruction-level behavior of the processor while abstracting out implementation details. It is then used as a reference against which the detailed design is compared. This high-level model is often referred to as an ISA (Instruction Set Architecture) or ISS (Instruction Set Simulator) model.

Because of their high level of abstraction, ISA models allow for very fast simulation. Software developers can use the ISA model stand-alone for developing initial versions of the system software.

For hardware/software cosimulation, the ISA needs to be connected to the RTL simulator. Typically a BFM provides this interface, as shown in Figure 8-3 on page 196. This BFM takes read and write commands from the ISA model and translates them into transactions on the bus. An additional piece of software (the API in Figure 8-3) may be used to interface the model to the BFM and to the hardware/software cosimulation environment.

One way to increase the performance of hardware/software cosimulation is to provide a model for system memory in the cosimulation environment rather than in the RTL for the design. Most of the bus accesses for a processor are memory accesses. If the model does not have to go through the BFM to RTL simulation to read/write system memory, then many RTL simulation cycles, and hence much simulation time, can be saved. Memory optimizations like this are one key advantage of having an API from the ISA model to the cosimulation environment.

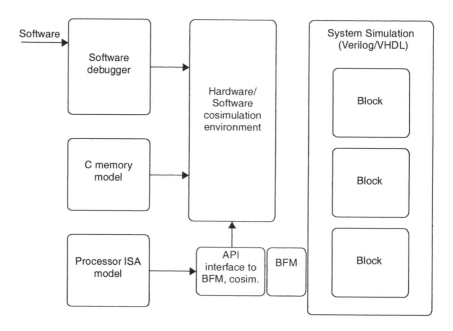

Figure 8-3   Hardware/software cosimulation using an ISA model

Behavioral models are the equivalent of ISA models for non-processor designs. Behavioral models represent the algorithmic behavior of the design at a very high level of abstraction, allowing very high speed simulation. For example, for a design using an MPEG macro, using a behavioral model instead of an RTL model can provide orders of magnitude faster system simulation.

A representative flow for compiling the behavioral models is shown in Figure 8-4.

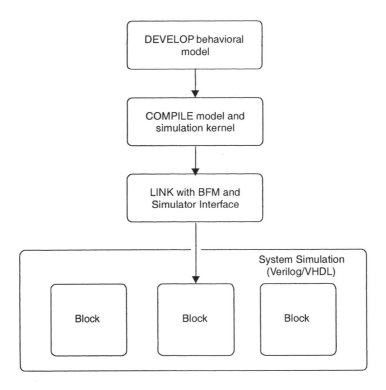

Figure 8-4   Generating compiled behavioral models

Cycle-Accurate Models

Cycle-accurate models are models that accurately reflect the behavior of the macro on a cycle-by-cycle basis. They are slower but more accurate than ISA models; as a result, they are more useful for hardware verification and less useful for software development and debug.

Like ISA models, cycle-accurate models typically use a BFM to interface to the RTL simulator.

Cycle-accurate models operate in zero-delay mode, with zero delays on outputs and zero hold time requirements on inputs. This model can be used by the integrator as part of the RTL simulation of the chip.

Sign-Off Models

Final or sign-off verification of a chip design requires a sign-off model of the hard macro. This sign-off model will typically reflect the complete, detailed functionality of the macro as well as detailed, accurate timing.

Most hard macro design teams create the sign-off model from the gate-level netlist for the design. The netlist should have extracted SDF timing values back-annotated onto it.

Besides the additional timing information, sign-off models may be functionally different from cycle-accurate models. The sign-off model must model the test structures for the macro, typically scan chains. Most cycle-accurate models do not include this level of functional detail.

Emulation Models

One of the major problems with full functional models, like cycle-accurate and sign-off models, is the slow simulation speeds achieved with them. Emulation is one approach to addressing the problem of slow system-level simulation with full functional models.

Emulation requires that the model for the macro be compiled into a gate-level representation. We can provide the RTL directly to the integrator, who can then use the emulator's compiler to generate the netlist, but this does not provide any security.

An alternate approach is to provide a netlist to the integrator. This approach provides some additional security for the macro. A separate synthesis of the macro, compiling for area with no timing constraints, will give a reasonable netlist for emulation without providing a netlist that meets the full performance of the macro.

Some emulation systems have more sophisticated approaches to providing security for hard macro models. See Chapter 11 for a brief discussion on this subject.

Hardware Models

Hardware models provide an alternate approach for providing highly secure full functional models. Because the hard macro design process requires that we produce a working test chip for the macro, this approach is often a practical form of model generation.

Hardware modelers are systems that allow a physical device to interface directly to a software simulator. The modeler is, in effect, a small tester that mounts the chip on a small board. When the pins of the device are driven by the software simulator, the appropriate values are driven to the physical chip. Similarly, when the outputs of the chip change, these changes are propagated to the software simulator.

Rapid prototyping systems also allow a physical chip to be used in modeling the over-all system. These systems are described in Chapter 11.

Some emulators allow physical chips to be used to model part of the system. Thus, the test chip itself is an important full functional model for the macro.

In all these cases, it is important that the physical chip reflect exactly the functionality of the macro. For example, with a microprocessor, one might be tempted to make the data bus bidirectional on the chip, to save pins, even though the macro uses unidirectional data buses. This approach makes it much more difficult to control the core and verify system functionality with a hardware modeler or emulator.

8.5.2   Timing Models

Figure 8-5 on page 200 shows the process for developing a static timing analysis model for the hard macro. From the SDF back-annotated netlist for the macro, the timing analysis tool extracts a black-box timing model for the macro. This model provides the setup and hold time requirements for input pins, the pin capacitance for input pins, and the clock-to-output delays for the output pins. This model is delivered as the standard Liberty format. During static timing analysis on the entire chip, the static timing analysis tool uses the context information, including actual ramp rates and output loading, to adjust the timing of the hard macro model to reflect the macro's actual timing in the chip.

For this black-box model to work, of course, the design must have no state-dependent timing. For blocks that do have state-dependent timing, a gray box timing model must be used; this model retains all of the internal timing information in the design. The entire back-annotated netlist can be used as a gray-box model, but it will result in slower static timing analysis runtimes.

One of the key problems in developing timing models is to provide useful and appropriate information about clock insertion delay. There are two possible usage scenarios for the hard macro: the integrator can provide a clock that arrives at the same time as the clocks for the rest of the chip, or he can provide an early clock. In the case of the early clock, the timing model must reflect the delay of the clock buffers in the hard macro. In the other case, the timing model must treat the clock insertion delay of the hard macro as zero.

We recommend that two timing models be developed, one with and one without the insertion delay. An alternative is to provide one model, but with a way to toggle the clock insertion delay between its actual value and zero. It is critical that the macro provider include a careful description of how to use the model in the user documentation. Any errors in dealing with the macro's clock delays could prove disastrous to the timing of the overall chip.

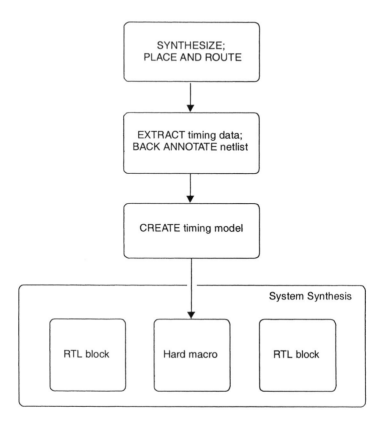

Figure 8-5   Generating static timing models for standard cell designs

8.5.3   Power Models

Power modeling for hard macros is still an immature technology. Accurate power modeling requires a detailed netlist stimulated with vectors that accurately represent a typical level of activity. These vectors are highly application-dependent; thus, any power model other than the full netlist is not very useful for analyzing power.

Most hard macro designers provide a rough estimate for the power of the macro in terms of milliwatts per megahertz (of the clock frequency), for different power supply values. Rather than trying to develop a reference set of vectors for determining this value, most macro providers base the value on the assumption that about 10% of the gates in a design will toggle at each clock.

8.5.4   Test Models

Because the hard macro appears to the chip designer as a black box, the macro designer must provide a test model for the macro. This model allows the chip designer to integrate the manufacturing tests for the macro, typically scan tests, into the overall manufacturing tests for the chip. This integration involves testing both the internals of the macro and the external logic adjacent to the macro.

The IEEE is currently defining a Core Test Language (CTL) [1] that defines a core (hard macro) test interface between an embedded hard macro and the system chip. The language facilitates reuse for embedded cores through core access and isolation mechanisms and provides testability for system chip interconnect and logic. In addition, the language supports core test interoperability with plug and play protocols. CTL supports many different forms of test (Scan, Iddq, BIST etc.) and does not dictate the test method employed within the core itself.

The macro test model is a CTL description of the test structures and strategies for the core, along with the scan vectors for the core.

The test model contains information on the testability and the test structures of the core, including:

- Clocks and synchronous signals
- Scan ports
- Scan chain specifications
- Test mode signals

The test model also captures the intent of all test modes and generates a test protocol for each mode. These modes include:

- Normal: This is the functional mode.
- Internal Scan: This mode is used to enable the internal scan chain(s) within the core.
- InTest: This mode tests the interface logic within the core. This is the logic between the IO port and a register, logic that is not tested typically by the internal scan.
- ExTest: This mode tests the logic external to the core and will place the wrapper into a state where the logic in the shadow of the core can be tested completely.
- Safe: This places the core into SAFE mode. In this mode, outputs are held constant, inputs ignored, and internal state preserved. Thus, the core will have no effect on scan tests external to it.

Test synthesis tools extract this information from the netlist of the core automatically, and create a test model that can be used in the test synthesis of the chip. The key to test model generation is to modify the macro design to make it fully test ready.

The first step is to use test synthesis to check the design for testability. Once any testability problems are fixed, the synthesis tool is used to insert scan chains in the design.

The rest of the process of creating a test ready core involves:

- Wrapping the core with a test wrapper
- Performing Multi-mode DRC on the wrapped core
- Inferring a CTL description for the wrapped core
- Creating test patterns for the wrapped core

The test synthesis tool can be used to generate the test wrapper. Since the wrapper cells are added to the design as part of a synthesis process, the impact of the cells on timing and area is considered and optimized to meet the timing constraints of the design. An example wrapper cell is shown in Figure 8-6 on page 203.

Once the test wrapper is generated, Multi-mode Design Rule Checking (DRC) validates the test model against the gate-level netlist and extracts the scan chains for each test mode.

The CTL description of the wrapped core is then created. This CTL description is an ASCII based model that adheres to an industry standard (IEEE P1500). As such, this model is fully portable among EDA tools that support P1500.

Finally, using Automatic Test Pattern Generation tools, the designer creates the test patterns for the macro.

The CTL description of the core, together with the test patterns for the core (ATPG) are deployed as the test model into the system chip level design environment.

Normal Mode

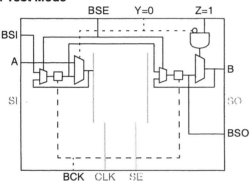

External Test Mode

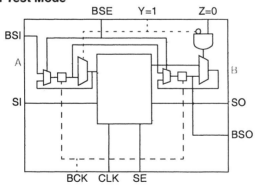

Internal Test Mode

Pin descriptions:

General:
- A: Functional input
- B: Functional output
- Y: Initialization value
- Z: Initialization value

External logic scan:
- BCK: Scan clock
- BSE: Scan enable (on logic 0)
- BSI: Scan input
- BSO: Scan output

Internal (core) scan:
- CLK: Scan clock
- SE: Scan enable
- SI: Scan input
- SO: Scan output

Figure 8-6 Modes of operation of wrapped core[1]

1. Originally published in "CTL the Language for Describing Core-Based Test" by R. Kapur et al., *Proceedings of the International Test Conference, 2001.* © 2001 IEEE.

8.5.5   Physical Models

What physical models need to be delivered with a hard macro depends on who ends up doing the physical design. In a typical ASIC environment, the integrator or chip designer may do an initial floorplan, and the rest of the physical design is done by the ASIC vendor. In other scenarios, the chip designer may do the complete physical design of the chip except for integrating the GDSII for the hard macro, which is treated as a block box. The chip designer then hands off the design to the silicon provider, who integrates the hard macro. Finally, in a semiconductor company, the chip designer may have access to all the view of the hard macros, and be responsible for the entire physical design.

In any of these scenarios, the hard macro designer must provide a floorplanning model of the macro. From the final place and route of the macro, the macro designer can extract the basic blockage information, pin locations, and pin layers of the macro. This information is typically delivered in the LEF format, and can be used in floorplanning the rest of the chip.

In addition, an abstract GDSII for the macro should be provided for those chip designers who need to do DRC on their physical design, but who do not have access to the full GDSII. This abstract view provides enough information on the macro, including signal pins, power pins, metal obstructions, and antenna diodes for DRC on the full chip to pass.

Finally, the macro designer should provide an LVS netlist to enable customers to perform LVS on the full chip. If the full GDSII is given to chip designer, then give full netlist needs to be provided; if an abstract GDSII is provided, then the corresponding abstract netlist needs to be provided.

8.6   Porting Hard Macros

One of the challenges for the hard macro provider is to port the macro rapidly from one process to another.

For hard IP that was completely synthesized, the porting strategy is quite straightforward. The designer just resynthesizes, targeting the new technology library, and repeat the physical design and timing model generation. If the designer has saved the scripts and the floorplan from the initial physical design, and these scripts were written to be as technology-independent as possible, then this is a reasonably painless process.

For those sections of the design that are full-custom, we have the choice between manual porting and automated porting. Under certain circumstances, automatic porting tools can be effective. These tools operate at the polygon level, automatically mapping transistors and interconnect from one set of design rules to another, and shrinking the design as much as possible.

This polygon mapping works quite well on cell libraries, and reasonably well on small blocks, perhaps hundreds of gates. As the blocks get larger, the chances increase that we will run into problems that require significant manual intervention. These problems can slow down the porting process dramatically. This is another reason why we recommend reserving full-custom design only for the most critical subblocks of the design.

The problems typically encountered in automated porting include clocking and hold time problems. As the technology shrinks, circuits speed up, and the acceptable clock skew becomes smaller, and minimum delays from flop to flop become less. These problems can be difficult to resolve, and can require adding or resizing gates and buffers.

One of the time-consuming aspects of using the porting tools is establishing the corresponding design rules for the source and target technologies. One way of reducing the risk of porting problems with automatic porting tools is to use lambda rules in the initial full-custom design. With these rules, all the physical constraints of the process are described as multiples of lambda, a unit length representative of the technology. If the design constraints of both the source and target libraries are both described as lambda rules, and the original macro design complies with the source library lambda rules, then automatic mapping is significantly easier.

References

1. Kapur, R., Keller, B., Lousberg, M., Reuter, P., Taylor, T., and Kay, D. "CTL the Language for Describing Core-Based Test," *Proceedings of the International Test Conference, 2001*, pp. 131-139. Website: http://grouper.ieee.org/groups/ctl/itc2001-ctl.pdf

CHAPTER 9

Macro Deployment: Packaging for Reuse

This chapter discusses macro deployment issues, including deliverables for hard and soft macros and the importance of keeping a design archive. The topics are:

* Delivering the complete product
* The contents of the user guide

9.1 Delivering the Complete Product

Once a macro has been designed according to the guidelines detailed in the preceding chapters, the macro must be packaged and delivered to the customer. The packaging of the macro depends on how it will be delivered.

Soft macros require:

* The RTL code
* Support files
* Documentation

Hard macros require:

* GDSII
* A rich set of models for system integration
* Documentation support for integration into the final chip

In addition, all files associated with the development of the macro must be stored together in a design archive so that all necessary information is available when it is time to modify or upgrade the macro.

As described in Chapter 4, a physical prototype of the macro is built as part of the verification phase of the macro development. Many third-party vendors make this prototype available to customers either as a demonstration of capability or as an evaluation unit. Evaluation prototypes are particularly helpful with programmable macros like microcontrollers and microprocessors; application code can be developed and run on the prototype to verify functionality and performance.

9.1.1   Soft Macro Deliverables

Table 9-1 lists the deliverables for soft macros.

Table 9-1   Deliverables for soft macros

Group	Deliverables
Product files	• Synthesizable Verilog for the macro and its subblocks • Synthesizable VHDL for the macro and its subblocks • Application notes, including Verilog/VHDL design examples that instantiate the macro • Synthesis scripts and timing constraints • Scripts for scan insertion and ATPG • Reference technology library • Installation scripts
Verification files	• Bus functional models/monitors used in testbench • Testbench files, including representative verification tests
Documentation files	• User guide/functional specification • Datasheet
System Integration files	• Bus functional models of other system components • Emulation models as appropriate to the particular macro and/or its testbenches and BFMs • Recommendation of commercially available software required for hardware/software cosimulation and system integration, as appropriate for the particular macro

Product Files

In addition to the RTL in Verilog and VHDL, we must include the synthesis and installation scripts. We include the reference technology library so that the customer can synthesize the design and verify that the installation was completed successfully. In addition, providing the reference library, scripts, and verification environment allows the user the recreate the developer's environment. This allows the user to verify many of the claims of the developer, in terms of timing, power, area, and testability of the macro.

The reference technology library is also very helpful in identifying library problems in the integrator's environment. Synthesis libraries vary considerably. If the integrator encounters synthesis problems with the vendor's library, the integrator can synthesize exactly the same configuration with the same scripts using the technology library. This process helps the integrator identify whether the problem is in the macro (and its scripts) or in the vendor's technology library.

Application notes that show exactly how to instantiate the design are also useful. If the application notes are available in soft form, the integrator can cut and paste the instantiation example, avoiding typographical errors and ensuring correct port names.

Verification files

The entire verification environment, including any bus functional models, bus monitors, or other models, and some set of verification test cases are shipped with the product. The test cases that ship with the macro usually do not represent the full test suite used to verify the macro. Typically, a subset is shipped that is sufficient to ensure that the macro has been installed correctly at the integrator's site.

The bus functional models used to develop and verify the macro can be used by the integrator to create a testbench environment for the SoC chip. See Chapter 11 for more discussion on using bus functional models for system-level testing.

System Integration Files

Depending on the specific macro, there may be additional deliverables that are useful for the integrator.

For large macros, where simulation speed in the system environment may be an issue, it can be useful to include cycle-based simulation and/or emulation models. In general, RTL that complies with the coding guidelines in this document will work with cycle-based simulation and emulation. However, testbenches and bus functional models, unless coded to these same RTL guidelines, may not be usable with these verification tools. It is up to the macro provider to determine which models need to be provided in cycle-based simulation/emulation compatible forms.

For macros that have significant software requirements, such as microcontrollers and processors, it is useful to include a list of compilers, debuggers, and real-time operating systems that support the macro. For other designs, we may want to reference software drivers that are compatible with the design. In most cases, the macro provider will not be providing the software itself, but should provide information on how to obtain the required software from third-party providers.

9.1.2   Hard Macro Deliverables

Table 9-2 lists the deliverables for a hard macro.

The list of deliverables in Table 9-2 assumes that the physical integration is being done by the silicon vendor rather than by the chip designer who is using the macro. This model applies when the silicon vendor is also the macro vendor. In the case where the chip designer is also doing the physical integration of the macro onto the chip, the physical GDSII design files are also part of the deliverables. These GDSII deliverables for the macro include up to three different implementations: one with no external power ring, and (optionally) two with power rings in different orientations.

Table 9-2   Deliverables for hard macros

Group	Deliverables
Product files	• Installation scripts
Verification files	• None
Documentation files	• User guide/functional specification • Datasheet • Description of models included in deliverables
System Integration files	• ISA or behavioral model • Bus functional model for macro • Cycle-accurate model • Sign-off model • Emulation models as appropriate to the particular macro • Timing and synthesis model for macro • Floorplanning model for macro • Recommendation of commercially available software required for hardware/software cosimulation and system integration, as appropriate for the particular macro • Test patterns for manufacturing test, where applicable

The other deliverables for hard macros consist primarily of the documentation and models needed by the integrator to design and verify the rest of the system. These models are described in Chapter 8.

For processors, an ISA model provides a high level model that models the behavior of the processor instruction-by-instruction, but without modeling all of the implementation details of the design. This model provides a high speed model for system testing, especially for hardware/software cosimulation. Many microprocessor vendors also provide a tool for estimating code size and overall performance; such a tool can help determine key memory architecture features such as cache, RAM, and ROM size.

For other macros, a behavioral model provides the high speed system-level simulation model. The behavioral model models the functionality of the macro, on a transaction-by-transaction basis, but without all the implementation details. A behavioral model is most useful for large macros, where a full functional model is too slow for system-level verification.

For large macros, bus functional models provide the fastest simulation speed by modeling only the bus transactions of the macro. Such a model can be used to test that other blocks in the system respond correctly to the bus transactions generated by the macro.

The cycle-accurate and sign-off models for the macro allows the integrator to test the full functionality of the system, and thus is key to system-level verification.

As in the case of soft macros, emulation models, especially for the macro testbench, may be useful for the integrator. These models are optional deliverables.

The timing models provide the information needed by the integrator to synthesize the soft portion of the chip with the context information from the hard macro. These models provide the basic timing and loading characteristics of the macro's inputs and outputs.

The floorplanning model for macro provides information the integrator needs to develop a floorplan of the entire chip.

Test patterns for manufacturing test must be provided to the silicon manufacturer at least, if not to the end user. For scan-based designs, the ATPG patterns and control information needed to apply the test patterns must be provided. For non-scan designs, the test patterns and the information needed to apply the test patterns is required; usually access is provided through a JTAG boundary-scan ring around the macro.

Documentation

In addition to the requirements for a soft macro, the documentation for a hard macro must provide descriptions of the various models provided as deliverables. The documentation also includes information on where the following data can be found:

- Footprint and size of the macro
- Detailed timing and power specification
- Routing restrictions and porosity
- Power and ground interconnect guidelines
- Clock and reset timing guidelines, including clock tree insertion delay
- Test integration guidelines

9.1.3   Software

As SoC designs get more complex, software is becoming an increasingly important aspect of chip design. One useful definition of an SoC design is one with a processor, and hence software, on chip. As IP becomes more complex, it becomes increasingly important to provide some level of software with the IP.

For simple macros, it is reasonable to provide the software team with a register manual and functional description, and have them develop the software from scratch. For more complex IP, this approach is open to too much interpretation (and mis-interpretation).

For complex macros, it is necessary to provide at least the first level of software routines so that the software team can use them as the basis for writing the higher levels of software. For example, if the macro is a bus peripheral such as a timer, the IP provider should provide basic functions to read and write the timer registers, enable the timer, set the timer count value, and set up the timer mode. The software team can then call these functions to set up the timer for some specific application.

Because the IP provider usually does not know the application of the final chip, or even the application language, it may not be possible to productize these low-level functions completely. Instead, these low-level functions may best be presented as examples. They should typically be provided in C, Verilog, and VHDL.

For configurable IP, register addresses and data bus widths may be parameterized, making accurate examples a bit trickier to produce. In this case, the scripts or tools used to configure the macro should also configure the example code.

9.1.4   The Design Archive

Table 9-3 lists the items that must be stored together in the design archive. All of these items are needed when any change, upgrade, or modification is made to the macro. The use of a software revision control system for archiving each version is a crucial step in the design reuse workflow, and will save vast amounts of aggravation and frustration in the future.

Table 9-3   Contents of the design archive

Group	Contents
Product files	• Synthesizable Verilog for the macro and its subblocks
	• Synthesizable VHDL for the macro and its subblocks
	• Reference technology library
	• Verilog /VHDL design examples that instantiate the macro
	• Synthesis scripts
	• Installation scripts
Verification files	• Bus functional models/monitors used in testbench
	• Testbench files
Documentation files	• User guide/functional specification
	• Technical specification
	• Datasheet
	• Test plan
	• Simulation log files
	• Simulation coverage reports (VHDLCover, VeriSure, or equivalent)
	• Synthesis results for multiple technologies
	• Testability report
	• Lint report that demonstrates compliance to coding guidelines
System Integration files	• Bus functional models of other system components
	• Recommendation of commercially available software required for hardware/software cosimulation and system integration, as appropriate for the particular macro
	• Cycle-based simulator and hardware emulator models

9.2   Contents of the User Guide

The user guide is the key piece of documentation that guides the macro user through the selection, integration, and verification of the macro. It is essential that the user guide provides sufficient information, in sufficient detail, that a potential user can evaluate whether the macro is appropriate for the application. It must also provide all the information needed to integrate the macro into the overall chip design. The user guide should contain, at a minimum, the following information:

* Architecture and functional description
* Claims and assumptions
* Detailed description of I/O
* Exceptions to coding/design guidelines
* Block diagram
* Register map
* Timing diagrams
* Timing specifications and performance
* Power dissipation
* Size/gate count
* Test structures, testability, and test coverage
* Configuration information and parameters
* Recommended clocking and reset strategies
* Recommended software environment, including compilers and drivers
* Recommended system verification strategy
* Recommended test strategy
* Floorplanning guidelines
* Debug strategy, including in-circuit emulation and recommended debug tools
* Version history and known bugs

The user guide is an important element of the design-for-reuse process. Use it to note all information that future consumers of your macro need to know in order to use the macro effectively. The following categories are especially important:

Claims and assumptions

Before purchasing a macro, the user must be able to evaluate its applicability to the end design. To facilitate this evaluation, the user guide must explicitly list all of the key features of the design, including timing performance, size, and power requirements. If the macro implements a standard (for example, the IEEE 1394 interface), then its compliance must be stated, along with any exceptions or areas where the macro is not fully compliant to the published specification. VSIA suggests that, in addition to this information, the macro documentation include a section describing how the user can duplicate the development environment and verify the claims.

For soft IP, the deliverables include a reference library, complete scripts, and a verification environment, so these claims can be easily verified.

For hard IP, the end user does not have access to the GDSII, and so many of the claims are unverifiable. We recommend including actual measured values for timing performance and power in the user guide.

Exceptions to the coding/design guidelines

Any exceptions to the design and coding guidelines outlined in this manual must be noted in the user guide. It is especially important to explain any asynchronous circuits, combinational inputs, and combinational outputs.

Timing specifications and performance

Timing specifications include input setup and hold times for all input and I/O pins and clock-to-output delays for all output pins. Timing specifications for any combinational inputs/outputs must be clearly documented in the user guide. Timing for soft macros must be specified for a representative process.

CHAPTER 10

System Integration with Reusable Macros

This chapter discusses the process of integrating completed macros into the whole chip environment. The topics are:

- Integration overview
- Integrating soft macros
- Integrating hard macros
- Integrating RAMs and datapath generators
- Physical design

10.1 Integration Overview

Chapter 2 described system design from specification to the point where individual blocks could be designed. The succeeding chapters described how these blocks should be designed in order to make them reusable. We now return to the issue of system design, and discuss how to assemble these blocks into the final chip.

At this point in system design, there are two key tasks remaining: physical design and functional verification. Each of these tasks has a dominant challenge. For physical design it is achieving timing closure; for verification, it is knowing when we are done, when we are confident enough in the functionality of the chip that we can tape out and go to fabrication.

In this chapter, we address the integration of the blocks and the physical design of the chip. In the next chapter, we discuss functional verification.

The process of integrating the blocks and doing the physical design can be broken into the following steps:

- Selecting IP blocks and preparing them for integration
- Integrating all the blocks into the top-level RTL
- Floorplanning and timing model generation
- Synthesis and initial timing analysis
- Physical synthesis and timing analysis, with iteration until timing closure
- Detailed route, timing verification, and power analysis
- Physical verification of the design

10.2 Integrating Macros into an SoC Design

Integrating macros into the top-level SoC design poses several challenges. In this section, we discuss typical integration problems and strategies for dealing with them.

10.2.1 Problems in Integrating IP

Assembling a set of blocks into a top-level design presents a series of challenges to the design team. Naturally, we did a good job of decomposing the design into well-specified blocks, then selected the IP we needed and designed the new blocks required as specified. Nonetheless, when we get down to assembling these blocks and making them work together, we often find issues.

For blocks that were designed specifically for this chip, we tend to find:

- The low-level interfaces do not work; for example, a handshake signal is inverted.
- There was a misunderstanding of the functionality of the block.
- There are functional bugs in the design.

Usually we have access to the block designers and the system architect, so these problems are reasonably easy to fix.

For IP that has been obtained from an external source, either a third party or some other division of the company, there are additional problems that frequently occur:

- Someone on the team needs to become familiar enough with the IP to integrate it into the design.
- The documentation is incomplete, making this understanding harder to obtain.
- The interface of the IP does not match the interface of the system bus.
- The verification models, such as bus functional models, are poorly written and difficult to integrate into the system verification environment.
- Only limited support is available from the IP provider.

We will defer the discussion of the verification issues until the next chapter. For now we will focus on the most serious of the other problems: interfaces that don't match the system.

It is not unusual for a team to purchase a piece of IP that consists of 20K gates or so, and then find that they have to design an additional block of 20K gates just to interface it to the rest of the system. Most digital block interfaces are designed to pass data; that is, they perform data reads and writes to other blocks. The protocol for these transactions may be quite different between different designs, and differ at different levels.

The detailed handshake may differ; one block may required a "ready for data" signal from the target before it does a write, while the target may expect a "request for write" signal before it reports status. At a higher level, blocks may have different kinds of transactions: posted writes, burst reads with or without out of order return data, and interrupted or aborted transactions. Incompatibilities at this level are more difficult to resolve.

The most difficult interface problems usually involve exception handling: interrupts, aborted transactions, and other unusual transactions. Differences at this level may have to be resolved at a high level, perhaps even requiring changes to the architecture of one of the blocks or the entire system.

10.2.2 Strategies for Managing Interfacing Issues

There is no universal solution for these interfacing issues except to adopt a universal interface standard. Some groups are attempting to establish internal standards within their companies, but we are a long way from having anything approaching a uniform standard across the industry. The power, performance, and protocol needs of different designs are just too disparate to make this approach practical.

There are several steps designs teams can take, however, to mitigate the problem:

- Plan the interfaces. We can identify early the kinds external IP to be used and analyze the interface protocols involved. We can then select the specific IP, and define the interfaces for the custom blocks, so that they can all be integrated with a minimum amount of additional interface design. What additional interface design is required can be included in the overall project plan. The main idea here is not to leave these interface issues until the last moment, and then be surprised at the additional work, and schedule slip, involved.

- Keep all interfaces as simple as possible, whether we are designing IP or custom blocks. These interfaces should usually include data read and ready for data signals, so the connecting blocks know the ready status at all times.

- Standardize on industry-standard buses.

- Accumulate IP and experience with the IP. Once a team has gained experience with a piece of IP, has used it successfully in a design, and has learned how to interface it to other blocks, that IP has significantly increased in value. There is a significant advantage to building a library of such IP, and leveraging it to create new designs. Some software reuse books talk of "product line planning", where multiple related products are developed over time to leverage investments in reusable IP.

- Document this expertise. If a piece of IP has deficient documentation, supplement it with the knowledge accumulated using it in a design. One of the most challenging aspects of using someone else's design is learning how it works, and how to use it. Capturing this knowledge in a document can help other integrators of the IP.

10.2.3   Interfacing Hard Macros to the Rest of the Design

In addition to the issues discussed above, hard IP presents some additional challenges for the integrator. For soft macros, power and clock tree routing, as well as scan insertion, are done during chip-level integration. This fact ensures the consistent and compatible power, clock, and test structure design. For hard macros, these are done during macro design, and the interfaces between the hard macro and the rest of the chip must be well thought out before integrating the macro into the chip.

Power and ground

Typically, the macro also has its own power and ground rings within the macro. The physical design of the rest of the chip must account for this, and provide the appropriate power and ground connections to the macro.

Test structures

Well-designed hard macros have their own embedded testability structures. These may include a JTAG port or a full-scan port. The macro may also have embedded structures for facilitating debug. These structures must be integrated into the overall chip design.

Clock distribution

Typically, the macro has its own internal clock tree. The overall clock distribution for the chip must accommodate the (already fixed) timing of the hard macro clock. In some cases, a clock output from the hard macro is used to synchronize the clocks for the rest of the system.

Guideline – The hard macro clock pin should be connected to a separate clock net in the design. The clock source could be the same, but the net has to be different. This will allow clock delay tuning and matching.

10.3 Selecting IP

In addressing the issues raised in the previous sections, one key step is to select IP that can be easily integrated into the overall chip design. Choosing well-designed, well-documented IP can greatly reduce the integration effort.

10.3.1 Hard Macro Selection

The first step in selecting a macro from an external source, or in specifying a macro that is to be developed by an internal source, is to determine the exact requirements for the macro. For microprocessor cores, this means developing an understanding of the instruction set, interfaces, and available peripherals.

Once the requirements for the macro are fully understood, the most critical factors affecting the choice between several competing sources for a hard macro are:

Quality of the documentation

Good documentation is key to determining the appropriateness of a particular macro for a particular application. The basic functionality, interface definitions, timing, and how to integrate and verify the macro should be clearly documented.

Completeness of the design and verification environment

In particular, functional, timing, synthesis, and floorplanning models must be provided.

If the macro is a microprocessor core, the vendor should supply or recommend a third-party supplier for the compilers and debuggers required to make the system design successful.

Robustness of the design

The design must have been proven in silicon.

Physical design limitations

Aspect ratio, blockage and porosity of the macro—the degree to which the macro forces signal routing around rather than through the macro—must be considered. A design that uses many macros that completely block routing may result in very long wires between blocks, producing unacceptable delays.

10.3.2 Soft Macro Selection

The first step in selecting a macro from an external source, or in specifying a macro that is to be developed by an internal source, is to determine the exact requirements for the macro. For a standards-based macro, such as a PCI core or a IEEE 1394 core, this means developing a sufficient understanding of the standard involved.

Once the requirements for the macro are fully understood, the choices can quickly be narrowed to those that meet the functional, timing, area, and power requirements of the design. The most critical factors affecting the choice between several competing sources for a soft macro are:

Quality of the documentation

Good documentation is key to determining the appropriateness of a particular macro for a particular application. The basic functionality, interface definitions, timing, and how to configure and synthesize the macro should be clearly documented.

Robustness of the verification environment

Much of the value, and the development cost, of a macro lies in the verification suite. A rich set of models and monitors for generating stimulus to the macro and checking its behavior can make the overall chip verification much easier. These models and monitors should be compatible with the chip-level verification environment.

Robustness of the design

A robust, well-designed macro still requires some effort to integrate into a chip design. A poorly designed macro can create major problems and schedule delays. Verifying the robustness of a macro in advance of actually using it is difficult. A review of the deliverables for compliance to the design, coding, and verification guidelines in this book is a first step. But for a macro to be considered robust, it must have been proven in silicon.

Ease of use

In addition to the issues above, ease of use includes the ease of interfacing the macro to the rest of the design, as well as the quality and user-friendliness of the installation and synthesis scripts. Some IP providers offer user interface tools to make soft cores easier to use.

10.3.3 Soft Macro Installation

The macro, its documentation, and its full design verification environment should be installed and integrated into your design environment much like an internally developed block. In particular, all components of the macro package should be under revision control. Even if you do not have to modify the design, putting the design under revision control helps ensure that it will be archived along with the rest of the design, so that the entire chip development environment can be recreated if necessary.

10.3.4 Soft Macro Configuration

Many soft macros are configurable through parameter settings. Designing with a soft macro begins with setting the parameters and generating the complete RTL for the desired configuration. A key issue here is to make sure that the combination of parameter settings is consistent and correct. Some IP providers supply configuration wizards with their IP to guide the user and prevent illegal configurations of the IP.

10.3.5 Synthesis of Soft Macros

The final step in preparing the IP for integration is to perform an initial synthesis with the target technology library. This initial synthesis can give a good preliminary indication of whether the macro will meet the timing, area, and power goals of the design.

10.4 Integrating Memories

Memories are a special case of the hard macro, and are worth some additional comment.

Large, on-chip memories are typically output from memory compilers. These compilers produce the functional and timing models along with the physical design information required to fabricate the memory. The issues affecting memory design are identical to those affecting hard macro designs, with the following additional issues:

- The integrator typically has a wide choice of RAM configurations, such as single port or multi-port, fast or low-power, synchronous or asynchronous.

- Asynchronous RAMs present a problem because generating a write clock requires a very timing-critical design that is tricky to create and difficult to verify. A fully synchronous RAM is strongly preferred.

- Large RAMs with fixed aspect ratios can present significant blockage problems. Check with your RAM provider to see if the aspect ratio of the RAMs can be modified if necessary.

- BIST is available for many RAM designs, and can greatly reduce test time and eliminates the need to bring RAM I/O to the chip's pins. However, the integrator should be cautious because some BIST techniques do not test for data retention problems.

10.5 Physical Design

The major challenge in the physical implementation of large SoC design is achieving timing closure. This process is inherently iterative; a typical spiral process where each iteration gets us closer to our performance goals. The problem that design teams often encounter is that many iterations are required to achieve their timing objectives, and each iteration can take many days. The result is often major delays to the project.

Many of the design guidelines in this book are intended to minimize the number of iterations in physical design by making timing closure as contained and local a problem as possible. In particular, the rules on partitioning, registering outputs, and fully synchronous design are key to containing the timing closure problem.

In this section we outline a process that can help make the iterations as few and quick as possible.

Figure 10-1 on page 225 outlines the process of integrating the various blocks into the final version of the chip and getting the chip through physical design. There are several variations on the flow shown here depending on the size of chip, the number of hard and soft blocks, and targeted performance. We describe here a representative flow, and will discuss briefly some of the main variants.

This process consists of four major activities:

- **Preparation of the design** – Planning the physical implementation of the chip, performing block-level and then chip-level synthesis.
- **Physical placement** – Doing a detailed floorplanning and initial placement, analyzing the timing results, and modifying the placement until timing goals are met.
- **Timing closure** – Adding clock buffers and performing detailed routing, doing a more accurate timing analysis, and fixing any remaining timing problems.
- **Physical verification** – Running the final checks on the design prior to tapeout.

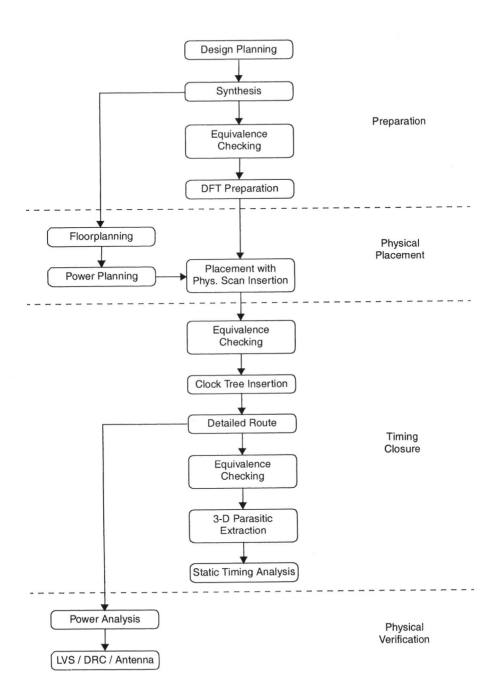

Figure 10-1 Integrating blocks into a chip

10.5.1   Design Planning and Synthesis

Preparation of the design involves a thorough understanding of the design flow and the gathering of necessary design libraries. It is a good practice for the design team to ensure that the starting RTL and corresponding libraries are tool-friendly.

Design Planning

Physical design starts with planning, and this planning can be done quite early in the design process. At the very start of the design, before blocks are designed or IP selected, the team should do an initial estimate of die size, number of I/O pads, and power dissipation. This information is key for determining package type.

Once the team has partitioned the design into blocks, the team can do a preliminary floorplan. This initial floorplan should include a rough placement of blocks and I/O pads, as well as some preliminary planning for the power and clock distribution. This information can be used to provide more accurate wire load models and timing budgets for synthesis.

If the inputs and outputs of each block are registered, then the timing budget is quite straightforward — the block just has to meet the clock frequency target of the design. The wire load model can be determined from the gate count of the block. The floorplanning information primarily helps identify long wires between blocks, or from blocks to I/O pads, which will required extra buffering.

If only the outputs of each block are registered, then the relative placement of the block on the chip affects both the wire load model and the time budget of the block. In Figure 10-2(a) on page 227, the blocks are close so that the arrival times at the inputs of Block B are nearly the same as the output time of Block A. The wire load model for Block A is probably accurate enough for the outputs of Block A as well as the internal signals. In Figure 10-2(b), the blocks are at opposite corners of the chip; this can mean a significant wire delay between the blocks. The outputs in Block A must be buffered up to drive the capacitance of the long wires, and the timing budget of Block B must be modified to allow for a later arrival time at its inputs.

Once the team has RTL for the soft blocks and physical libraries (LEF) for the hard blocks, the team can use a floorplanning tool to refine the floorplan. In particular, the team can assign physical locations for the I/Os of each block and do top-level routing. This approach can give very accurate estimates of the capacitive loading on the top-level interconnect, making synthesis much more accurate.

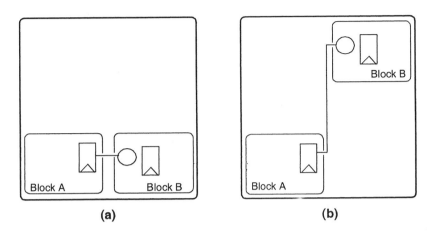

Figure 10-2   The floorplan affects timing budgets and wire load models

Synthesis

Armed with this preliminary physical design information, the design team can now do a full synthesis of the chip. Using the wire load models and timing budgets from the initial floorplan, we synthesize each block independently.

Once each block is meeting timing, we do a top-level synthesis of the entire chip, using timing models for the hard macros. At the top level, synthesis should be required only to stitch together the top-level netlist and refine the system-level inter-connect: resizing buffers driving inter-block signals, fixing hold-time problems, and the like. For critical inter-block paths, some re-budgeting may be required. For this, we can go back to the design planning tool to readjust block placement, or I/O place-ment for the key blocks, or to re-route some top-level nets. Then, we can generate a new top-level timing budget and wire loads, and re-run block-level synthesis.

The inputs to the top-level synthesis may include:

- Timing budgets and wire load models from the design planning stage
- RTL (or a netlist synthesized from RTL) for the synthesizable blocks
- Synthesis models for the hard macros and memories
- Netlists for any modules generated from a datapath generator
- Any libraries required, such as the DesignWare Foundation Library
- Top-level RTL that instantiates the blocks, the I/O pads, and top-level test struc-tures

The synthesis models for the hard macros and memories include the timing and area information required to complete synthesis on the whole chip and to verify timing at the chip level.

The top-level test structures typically include any test controllers, such as a JTAG TAP controller or a custom controller for scan and on-chip BIST (Built-In Self Test) structures.

If clock gating is needed to reduce power, then the power compiler should be used to convert mux-hold flops into gated clocks.

Once all the test structures are in place, a final timing analysis is performed to verify chip timing and, if necessary, an incremental synthesis is performed to achieve the chip timing requirements.

The synthesized netlist, along with any relevant physical database, is now ready for detailed floorplanning.

Note: As we go through the block and chip-level budgeting and synthesis, we begin to realize the benefit of some of the design and coding guidelines. In particular, it quickly becomes obvious that false paths and timing exceptions present a real problem. Any exception to the basic timing goals, such as paths that take two cycles, or test signals that do not have to meet the operating frequency, need to be listed in the synthesis and budgeting scripts. This manual process is very prone to error. The authors have seen large chips where there were literally thousands of timing exceptions. In cases like these, the designers consistently miss a significant number of paths, either specifying a path as false when it is not, or the other way around.

Either of these cases can result in synthesis and timing problems. If the path is not false and we mark it as false, then clearly it will not be synthesized to meet timing. On the other hand, if a path is false and is not marked as false, then synthesis will work hard to get it to meet timing, often to the extent of not optimizing other paths that really are critical.

Thus, to meet timing, it is essential that the false path lists be completely correct. For this reason, we strongly recommend that designers avoid false and multicycle paths completely. In the worst case, the list of paths should be very short.

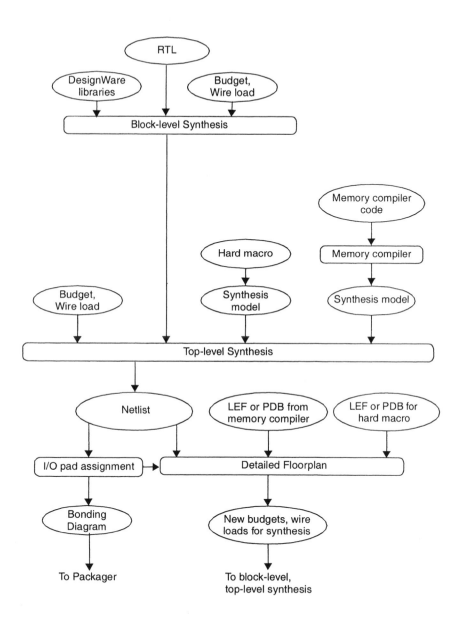

Figure 10-3   Chip-level synthesis

Equivalence Check

Functional verification tools can be used to prove or disprove the functional equivalence of two designs. Such tools can significantly reduce the design cycle by providing an alternative to simulation for regression testing. For example, you can use these tools to compare a synthesized netlist to its RTL source or to a modified version of that netlist. After the comparison, a report log should indicate whether the two design are functionally equivalent. Performing equivalence check ensures that the newly implemented design is functionally equivalent to the reference design. Doing so greatly reduces number of complications in future debugging and verification.

DFT Preparation

After the top-level netlist has been generated, scan cells should be inserted in the appropriate blocks for testability. Scan insertion is typically done by a DFT (Design for Testing) tool. Note that at this point, the DFT tool merely replaces standard flops with scan flops. It is the designer's responsibility to prepare the database to meet tool requirements, such as placed and appropriately declared scan ports and scan signals. DFT preparation also includes determining the scan methodology and tool configuration such as how to deal with mixed clocks or edges, if any. We recommend that the actual stitching of flops into the scan chain be performed with physical placement. Thus, the scan interconnect between flops, which is arbitrary, can be optimized for minimum wire length.

An ATPG (Automatic Test Pattern Generator) tool can then be used to generate scan test vectors for the chip.

10.5.2 Physical Placement

Physical placement is the critical stage in which timing closure is achieved. Synthesis can never have accurate enough wire-load models to account for physical placement information. In synthesis, all nets with the same fan-out have the same estimated interconnect delay. After placement, it is apparent that nets with the same fan-out will not have the same interconnect delay, and such estimation is the cause of inaccuracy of timing report.   This illustrates the importance of placement information in interconnect delay calculation.

Floorplanning and Power Planning

At this point, we can read the netlist into the floorplanning tool and complete the preparation for placement. We can fix, if we haven't already:

- Block placement
- I/O pad placement
- Placement of the I/O cells for each block

Next, we do an initial route of the power mesh, the distribution of power and ground in the chip. Designers have found that doing the power routing before detailed placement improves the overall design. Power routing takes routing resources on the chip that could otherwise be used for routing signals. We must account for the metal layers used in power distribution to ensure that there are still sufficient routing resources for signals. Depending on the number of available routing layers and signal congestion, it might be necessary to dedicate routing channels for critical nets. A prudently placed design should have timing-critical components placed away from congested area.

Typically power routing involves placing wide power and ground rings around the periphery of the chip, and then cross-hatching the chip with a mesh of power and ground wires. For large chips, the chip may be divided up into sections, with each section having its own power and ground rings.

Often, chips will have different power supplies (and power rings) for I/O and the core logic, especially if they run off of different voltages. For example, in low power designs, the core is often run at as low a voltage as possible (for example, 1.5v) while the I/O must run at standard voltages, such as 3.3v. Also, any analog block, such as a PLL or A/D converter, may need a separate power and ground supply to provide noise isolation.

Placement and Physical Scan Insertion

Once the synthesized netlist meets timing based on wire load models, the placement engine needs to place the design such that the timing goals of the design can be met. The effective use of timing-driven engines is the key to achieving this goal.

Timing-driven placement has been a goal of tool providers and engineers for many years. Today, physical synthesis provides a timing-driven placement technology that is mature enough to make timing closure on large chips dramatically easier than it has been in the past. To use this technology effectively, we need to provide to physical synthesis:

* A reliable technology file that accurately models the interconnect parasitics
* Reasonable and accurate timing constraints
* Comprehensive physical library for all cells undergoing placement
* A design which meets tool specifications and is optimization-friendly

Physical synthesis takes as input the timing constraints and the gate-level netlist (typically Verilog), as well as corresponding physical and logical libraries. It then attempts to place the cells in the design so as to meet the timing constraints, resynthesizing the design as necessary. The placement engine uses estimates for routing delays, so that a full route of the design is not required to determine if it meets timing. The accuracy of these estimates is key in achieving timing closure.

Physical synthesis relies on a technology file to tell it how to estimate the capacitance of metal interconnect. Coupling capacitance between adjacent wires (both beside the wire in question, and above and below it), greatly affect the total capacitance seen by the driving gate, and thus the delay of the gate. The capacitance model used by the placement engine must be pessimistic in order to ensure that estimated routing delays are no worse than those from the actual, final route. One way to do this is to force the tool to assume that each wire has other wires in adjacent routing tracks both beside and above and below the wire. The technology file is where we can specify this data to the placement engine.

In deep submicron designs, wires are taller than they are wide; as a result, coupling capacitance has a significant effect on overall capacitance. This effect must be modeled in the technology file to achieve accurate capacitance estimates.

Once the capacitance per unit length is well modeled, the placement engine must estimate the actual length of each interconnect. Most placement engines do this by assuming a Steiner route; that is, an optimal route based on orthogonal routing. Congested areas, though, may prevent some routes from being Steiner; they may have to take longer, "scenic" routes (like going from San Francisco to Cleveland by way of New York if all the direct flights are booked). The better placement tools must be able to estimate congesting and its effect on routing resources, and factor this into the delay estimate. Certain placement engines, use coarse route estimation, which for the most part is Steiner route but takes metal layer blockages into consideration. This introduces another degree of accuracy in modeling routing congestion.

Prior to scan insertion, we must ensure that the database is properly setup for scan circuitry. It is inherently advantageous to perform scan insertion during placement. DFT engine should work with the placement engine so to stitch and order the scan chain based on physical placement information. It has been shown that scan chains which undergo physical ordering greatly alleviate congestion and conserve routing resources.

After placement is complete, most placement engines generate a report of the estimated capacitance in the routing. This report is relatively quick to generate, but not as accurate as a full 3-D extraction. The accuracy depends largely on how the tool models parasitics.

Deriving Constraints

Well-derived constraints can make synthesis and timing-driven placement run faster and converge with fewer iterations. Deriving constraints for timing-driven placement is very much the same as synthesis. There are many factors which affects the result of timing-driven placement. Flop configuration is such an example. In particular, a fully synchronous, flop-based design can allow timing-driven placement to produce better results with less complications.

Accurate and appropriate constraints are essential to good timing-driven placement. The designer must consolidate as much performance specifications as possible and derive constraints accurately and appropriately based on those specifications. The designer must be able to discern whether a constraint is reasonable to avoid unwittingly and unnecessarily over-constraining the design.

It is critical to identify all false paths in a design prior to placement. If a false path is not identified, the placement engine can spend all of its time attempting to meet timing on this false path, and produce sub-optimal placement on actual critical parts of the design. Once again, avoiding false and multi-cycle paths can greatly help achieve rapid timing closure.

Flat vs. Hierarchical Placement

One critical issue in doing placement is deciding how much hierarchy to maintain during physical design.

Some designs (microprocessor designs, for example), are designed with a strict hierarchy that is maintained throughout physical design. Typically this includes:

- A careful floorplan is developed early, and a location for each major block identified.
- Pin locations for the I/O of each block are assigned.
- Some room between blocks is reserved for top-level routing; all routing between blocks is restricted to this area.
- Top-level routing is performed before place and route of the blocks; as a result, the wire length and capacitive loading for each top-level wire is fixed.
- Based on the information from the steps above, each block is placed and routed independently, and then placed in the top-level design.

A second approach is to do a completely flat place and route. In this approach:

- A careful floorplan is developed early, and a location for each major block identified.
- Based on the information from the steps above, timing constraints are developed for the design. Placement constraints are developed based on the floorplan, and the entire design is placed as a unit.
- Detailed routing is performed on the chip as a single unit.

The irony in the flat approach is that a detailed floorplan is still needed; it allows us to develop the timing constraints for placement.

Real designs may use a combination of the approaches above. Many teams will initially try a hierarchical approach. If the design still has problems meeting timing or has excessive routing congestion, they then will try a flat placement. Based on the results, they then pursue the approach that looks the most promising.

The only strong recommendation we make in this area is that the physical hierarchy should reflect the logical hierarchy. A physical block may consist of several logical blocks, but a single logical block should never be split across several physical blocks unless absolutely necessary to meet timing. The resulting name changes makes it very difficult to work with the post-layout netlist and to troubleshoot problems.

Timing Convergence and Iterations

One always hopes that after an initial placement, timing has been met and all that is required is to route the chip and tape out. One is almost always disappointed.

There are two major sources of timing problems at this point: the timing constraints, and the design itself.

If the design has false paths that are not listed in the constraints, then we are likely to find that the long paths in the design are paths we do not care about. But we are also likely to find that a number of critical paths were not appropriately placed, and are failing timing. The solution to this problem is to update the constraints and re-run physical synthesis.

If the timing is close but not quite passing, then it may be useful to refine the timing budgets. This can be done manually by changing the constraints or automatically by using a timing budgeting tool. Then we re-run placement.

If the timing is still not met, then we may have to modify the design itself, changing the RTL to add pipeline stages or the like. In this case, we have to repeat the floor-planning, synthesis, and initial placement.

Under any of the scenarios above, as well as a host of others, it becomes necessary to iterate through placement. The goal is to make this iteration as short as possible, so that we can converge quickly to a placement that meets timing. If our routing delay estimates are accurate, then we can then achieve full timing closure quickly.

10.5.3 Timing Closure

After placement meets timing, we have several key tasks to complete the design.

Clock Tree Insertion

Before doing a full route of the design, we must balance the insertion delay from the clock source to leaf cells by buffering the clock(s). This is also known as clock tree insertion. Since these are the most critical nets in the design, and need to be balanced to minimize clock skew, they are routed first. One common problem with buffering clock trees is that they sometimes require a very large number of buffers. For this reason, some designers reserve a buffer site near each group of flops. If the site is not needed, it takes up some small incremental area, but this is well worth it if it speeds convergence of clock tree insertion.

Physical synthesis tools can do clock tree budgeting while doing placement. Optimization after clock tree insertion is also possible.

Detailed Route

After the clock is routed and meeting skew, we do a full detailed route of the design. This is the first time we have a complete physical design. We can now do a much more accurate assessment of the timing and power.

During detailed route, we need to ensure that we comply with the process' rules for antennas. During chip fabrication there is a time during which a metal layer has been added to the chip, but the other metal layers have not. At this time, metal stubs in that layer can act like as antennas, picking up a static charge and damaging the chip. Each process has a set of rules for how long the stub can be for each metal layer. By adding these rules to the cell library, we can get the router to comply with them during route. Otherwise, it is necessary to go back after the route and fix any antenna violations, a time consuming process.

Parasitic Extraction

We now use a full 3-D extraction engine to calculate the actual load of each metal interconnect segment in the design. These extraction engines are full field solvers that give much more accurate delay modeling as compared to statistical interconnect load models used in synthesis and placement.

It is sufficient to use a statistical wire load model to model interconnect delays for older technologies where gate delay is the predominant contributor of delay. Interconnect capacitances of older technologies are primarily determined by area capacitance (wire-to-substrate capacitance) and fringing capacitance (wire side to substrate). The edges of two adjacent wires are so far apart that the capacitance from one wire to another can be ignored.

Recent deep sub-micron technologies have thinner metal lines and less space between adjacent wires. The older wire-load model failed in delay estimation for submicron processes due to the emerging dominance of interconnect delay. For the same length of wire, interconnect delays are much larger for two reasons: wires are narrower, and wires are closer together. Narrower wires increase the resistance of wires, while closer wires increase the coupling capacitance and cause cross-talk effects on signal nets.

Interconnect capacitances of submicron technologies are primarily determined by sidewall (or coupling) capacitance.With multi-metal process, capacitances need to be addressed in three dimensions. The different capacitances include wire-to-wire (same layer), wire-to-substrate, and wire-to-wire (different layers). Thus, it becomes increasingly crucial and necessary to have a reliable 3-D extraction engine to accurately model the timing of the design.

With the extracted data, we can now do a full static timing analysis and determine the timing of the design.

Static Timing Analysis

Static timing analysis tools can read these capacitances, along with the netlist and timing constraints, and output a timing report. This timing report lists all the paths that are violating the timing constraints (as well as a host of other reports, as required).

We then analyze the timing reports to determine if the violating paths are real and if so, what to do about them. If the violations are false paths, we update the timing constraints.

If the timing violations are real, then most of them will probably be from excessive capacitance loading gate outputs. The solution here is to increase drive strengths, add buffers, or even restructure logic.

Physical synthesis tools have the capability of doing much of this automatically as part of timing-driven placement. They can resize and/or add buffers to improve timing.

Timing Fixes

If the timing estimates used in placement were accurate, there should be few timing violations at this point. We would typically expect a couple of long paths that need repair. We also would expect some hold time violations. Both of these can result from the fact that we used estimated routing models. There may also be some remaining clock tuning required to meet our skew requirements.

We fix the clock and the long paths first; if there are literally just a couple of fixes required, we may be able to do these interactively in the place and route tool. For larger numbers of fixes, we may have to go back and readjust our timing constraints, re-optimize, and go back through place and route.

After these fixes have been implemented in the physical design, we again do a full extraction and timing analysis. We iterate as required until the clock meets our requirements.

Once the clock tree is finalized and is meeting timing, we need to fix any remaining hold time problems. Hold time problems result from a combination of fast data paths from register to register and clock skew. They are typically fixed by inserting buffers in the fast data paths.

Virtually all hold time problems should be fixed during placement. Hold time violations, like long path problems, are fixed by the in-place optimization process. A few new hold time problems may appear as the clock is tuned; these we fix at this point in the process.

10.5.4 Verifying the Physical Design

The last major step in the physical design process is verifying that the physical design is correct and in compliance with the design rules for the target silicon process.

Checking Power

First we do a check of the power distribution system. We can estimate the voltage drop across the power meshes using power analysis tools. Our initial power mesh design was intended to be conservative, so we should see no surprises here.

We can also use power estimation tools to get a final estimation of the power dissipation of the design.

DRC, LVS, and Antenna

Finally we run DRC (Design Rule Checking) and LVS (Layout vs. Schematic). DRC verifies that the design does not violate any physical design rules.

For full-custom designs, this step can involve many iterations as subtle problems with the placement of cells are discovered and fixed. But in the standard model of reuse presented in this book, full-custom designs should only be imported into SoC designs after they have been physically designed and verified. No full-custom DRC violations should occur at the chip level.

For standard cell designs, there should be very few DRC violations. Typical problems that do occur are usually caused by problems in the library or by interface problems between the standard cell sections and any hard blocks that have been imported. These problems are usually quite straightforward to fix.

LVS compares the design as physically implemented to the gate-level netlist. It extracts a post-layout netlist back from the physical design by mapping polygons back into gates. It then compares this post-layout netlist to the pre-layout netlist. Again, for standard cell designs, there tend to be few LVS errors in the final design. The ones that do appear tend to be library problems and are usually straightforward to fix.

Finally, we must check for any antenna violations that might have escaped antenna autofix during detailed route. We must manually amend these violations, either by tying the net to a diode or shorten the metal stub in a layout editor.

Of course, we need to make sure that the final netlist, with added buffers and resized gates, and clock fixes, still is functionally equivalent to the original netlist. Formal verification should be used to check this equivalence.

Once these steps are completed, the chip is ready for fabrication.

10.5.5 Summary

The physical design of very large chips is an extremely challenging and complex task. The algorithms used by the tools are very complex, and the databases huge. It is very easy to spend many months trying to reach timing closure for a large chip.

There is much the designer can do to reduce the risk of runaway schedules in physical design. The key is to make timing closure and physical design a series of local, relatively small problems. The process described above performs most of the real effort in timing closure during placement. Once placement is successful, the rest of the design process is straightforward and should require few iterations. The runtimes for extraction can be very long, and DRC and LVS can take several days. However, by ensuring a high probability of needing only one or two runs of each, this long runtime is tolerable.

The highly iterative loop is in timing-driven placement. By carefully choosing a set of simplifying assumptions, mainly in how we estimate routing delay, this loop can be made relatively fast (hours instead of days), so that we can tolerate these iterations.

Above all else, the most important key to rapid timing closure is the quality of the design itself. A fully synchronous design, with few or no timing exceptions, where the levels of logic between registers is well understood and consistent with the timing goals, can make it through physical design with few schedule surprises.

CHAPTER 11

System-Level Verification Issues

This chapter discusses system-level verification, focusing on the issues and opportunities that arise when macros are integrated into a complete System-on-Chip. The topics are:

- The importance of verification
- Test plan
- Application-based verification
- Fast prototype testing
- Gate-level verification
- Verification tools
- Specialized hardware for system verification

11.1 The Importance of Verification

Verifying functionality and timing at the system level is probably the most difficult and important aspect of SoC design. It is the last opportunity to find conceptual, functional, and implementation errors before the design is committed to silicon. For many teams, verification takes 50%-80% of the overall design effort.

For SoC design, verification must be an integral part of the design process from the start, along with synthesis, system software, bringup, and debug strategies. It cannot be an afterthought to the design process.

System verification begins during system specification. The system functional specification describes the basic test plan, including the criteria for completion (what tests must run before taping out). As the system-level behavioral model is developed, a testbench and test suite are developed to verify the model. Similarly, system software is developed and tested using the behavioral model rather than waiting for real hardware. As a result, a rich set of test suites and test software, including actual application code, should be available by the time the RTL and functional models for the entire chip are assembled and the chip is ready for verification.

Successful (and rapid) system-level verification depends on the following factors:

- Quality of the verification plan
- Quality and abstraction level of the models and testbenches used
- Quality and performance of the verification tools
- Robustness of the individual pre-designed blocks

11.2 The Verification Strategy

The system-level verification strategy for an SoC design uses a divide-and-conquer approach based on the system hierarchy. This strategy consists of the following steps:

- Verify that the leaf nodes — the lowest-level individual blocks — of the design hierarchy are functionally correct as stand-alone units.
- Verify that the interfaces between blocks are functionally correct, first in terms of the transaction types and then in terms of data content.
- Run a set of increasingly complex applications on the full chip.
- Prototype the full chip and run a full set of application software for final verification.
- Decide when it is appropriate to release the chip to production.

Block-Level Verification

For large SoC designs, it is essential that each block be fully verified before it is integrated into the chip design. In this sense, block-level verification is a prerequisite and precursor to chip-level verification.

Block-level verification is described in detail in Chapter 7. It uses code coverage tools and a rigorous methodology to verify the RTL version of macro as thoroughly as possible. A physical prototype is then built to provide silicon verification of functional correctness.

This verification methodology should, in general, be used for any block to be used in the chip design, even if that block is not intended for reuse. Verifying blocks fully before integration greatly reduces the overall verification effort, since bugs are much easier to find at the block level rather than chip level.

The only exception to this rule is that the design team may well decide not to produce prototypes of single-use blocks before they are integrated into the chip. This approach seems a reasonable risk/benefit tradeoff, but the risk involved should be recognized.

Any block in the SoC design that has not gone through this process, including silicon verification, is not considered fully verified as a standalone block. If the chip contains any such partially verified blocks, the first version of the chip must be considered a prototype. It is virtually assured of having bugs that require a redesign of the chip before release to production.

Prototyping the chip, however, is part of the overall chip verification plan, so it is reasonable to have some number of new, single-use blocks that have been robustly verified, but that have not been prototyped.

11.3 Interface Verification

Knowing that the individual blocks have been robustly verified, chip-level verification consists primarily of verifying the interfaces and interaction between the blocks. Thus we start chip verification with interface verification.

Inter-block interfaces usually have a regular structure, with address and data buses connecting the blocks and some form of control—perhaps a request/grant protocol or a request/busy protocol. The connections between blocks can be either point-to-point or on-chip buses.

Because of the regular structure of these interfaces, it is usually possible to talk about *transactions* between blocks. The idea is that there are only a few permitted sequences of control and data signals; these sequences are called transactions and only the data (and data-like fields, such as address) change from transaction to transaction.

11.3.1 Transaction Verification

Interface testing begins by listing all of the transaction types that can occur at each interface, and systematically testing each one. If the system design restricts transactions to a relatively small set of types, it is fairly easy to generate all possible transaction types and sequences of transaction types and to verify the correct operation of the interfaces to these transactions. Once this is done, all that remains is to test the behavior of the blocks to different data values in the transactions. Thus, a simple, regular communication architecture between blocks can greatly reduce the system verification effort.

In the past, this transaction checking has been done very informally by instantiating all the blocks in the top-level RTL, and then using a testbench to create activity within the blocks and thus transactions between blocks. If the overall behavior of the system is correct, perhaps as observed at the chip's primary I/O or in the memory contents, then the chip — and thus the interfaces — were considered to be working correctly.

There are several changes that can be made to improve the rigor of transaction checking. First of all, as shown in Figure 11-1(a) on page 243, you can add a bus monitor to check the transactions directly. This monitor can be coded behaviorally and thus provide very good simulation performance. For a chip like that shown in Figure 11-1(b), with point-to-point interconnections, it is possible to build some simple transaction checking into the interface module of each block. Testbench automation tools can be useful tools for creating effective transaction checkers very quickly.

This monitor approach improves observability during transaction testing, but it is also possible to improve controllability. If we use simple, transaction-generating bus functional models instead of full functional models for the system blocks, we can generate precisely the transactions we wish to test, in precisely the order we want. This approach can greatly reduce the difficulty of developing transaction verification tests and can reduce simulation runtime as well.

11.3.2   Data or Behavioral Verification

Once the transactions have been verified, it is necessary to verify that the block behaves correctly for all values of data and all sequences of data that it will receive in actual operation. In most chips, generating the complete set of these data is impossible because of the difficultly in controlling the data received by any one block.

The approach described above helps here as well. We use the bus functional models for all blocks except the block under test, for which we use the full RTL. We can then generate the desired data sequences and transaction from the BFMs. We can construct test cases either from our knowledge of the system or by random generation.

Automatic checking of the block's behavior under these sequences of transactions is nontrivial and depends on how easy it is to characterize the correct behavior of the block. For complex blocks, the semantics of a testbench generation tool may be the only way to describe the block's behavior such that its outputs can be checked automatically.

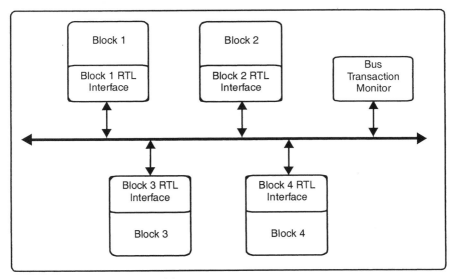

(a) Chip with an on-chip bus

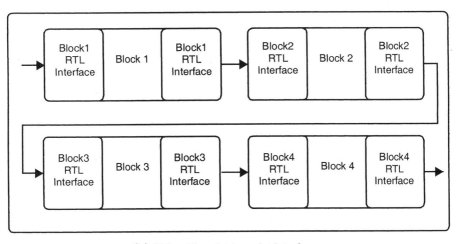

(b) Chip with point-to-point interfaces

Figure 11-1   System verification using interface testing

This test method often reveals that the block responds correctly to data sequences that the designer expected the block to receive, but that there are some (legal or illegal) sequences that can occur in the actual system to which the block does not respond correctly. This must usually be considered a bug, requiring redesign of the block.

Another method for dealing with the problem of unanticipated or illegal inputs is to design a checker into the block interface itself. This checker can suppress inputs that are not legal and prevent the block from getting into incorrect states. This approach has been used effectively in high-reliability system designs.

11.3.3 Standardized Interfaces

Interface verification and transaction checking can be greatly facilitated if the interfaces are standardized. Clearly it is easier to get a bus functional model or bus monitor out of a library than to create one from scratch.

This is one of several reasons why design teams are trying to standardize the primary interfaces to the chip and the on-chip buses. Once these standards are established, bus functional models and bus monitors can be developed and reused on many chip designs.

11.4 Functional Verification

Once the basic functionality of the system has been verified by the transaction testing, system verification consists of exercising the entire design, using a full functional model for most, if not all, of the blocks. The ultimate goal of this aspect of verification is to try to test the system as it will actually be used. That is, we come as close as we can to running actual applications on the system.

Verification based on running real application code is essential for achieving a high quality design. However, this form of verification presents some major challenges. Conventional simulation, even at the RTL level, is simply not fast enough to execute the millions of vectors required to run even the smallest fragments of application code, much less to boot an operating system or test a cellular phone.

There are two basic approaches to addressing this problem:

* Increase the level of abstraction so that software simulators running on workstations run faster.
* Use specialized hardware for performing verification, such as emulation or rapid-prototyping systems.

This section addresses the first approach: how to use abstraction and other mechanisms to speed conventional simulation techniques. Subsequent sections address the second approach.

The types of abstraction techniques we can use depend on the nature of the design, so it is useful to use a specific design as an example. Fortunately, most large chips are converging to an architecture that looks something like the chip design shown in Figure 11-2, the canonical SoC design described in Chapter 2.

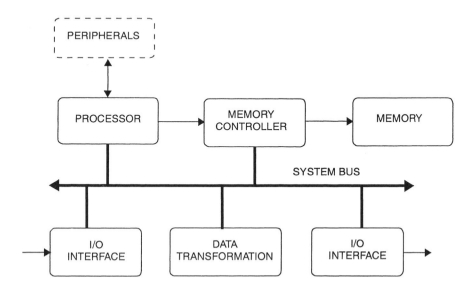

Figure 11-2 Canonical SoC Design

Figure 11-3 on page 246 shows a possible testbench environment for verifying the canonical design. The key features of this verification environment are:

- The full RTL model is used as the simulation model for most of the functional blocks.

- Behavioral or ISA (Instruction Set Architecture) models may be used for memory and the microprocessor.

- Bus functional models and monitors are used to generate and check transactions with the communication blocks.

- It is possible to generate real application code for the processor and run it on the simulation model.

With this test environment, we can run a set of increasingly complex application tests on the system. Initially, full functional models for the RAM and microprocessor are used to run some basic tests to prove that the system performs the most basic functions. The slow simulation speeds of this arrangement mean that we can do little more than check that the system is alive and find the most basic system bugs. At this level of abstraction, we are probably simulating at a rate of tens of system clocks per second.

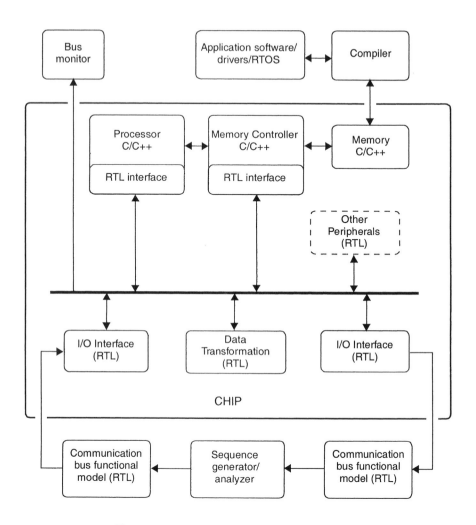

Figure 11-3   System verification environment

Behavioral models are now substituted for the memory and microprocessor. These models can be high-level C/C++ models that accurately model the instruction set of the processor, but abstract out all implementation detail. These models are often called ISA (Instruction Set Architecture) models. Another approach is to code a very high-level, behavioral model in Verilog or VHDL, abstracting out much of the cycle-by-cycle details, but retaining the basic functionality of the processor. If enough timing detail is retained so that the bus transactions at the I/O port of the processor are accurate on a clock-cycle by clock-cycle basis, the model is often referred to as a cycle-accurate model.

Using these behavioral models for the memory and processor, real code is compiled and loaded into the memory model and the processor model executes this code. At the same time, representative data transactions are generated at the communication interfaces of the chip, usually by bus functional models. For instance, if the data transformation block is an MPEG core, then we can feed in a digital video stream.

Using C/C++ models for both the processor and memory dramatically improves simulation speed over full RTL simulation. In designs like our canonical example, most cycles are spent entirely in the processor, executing instructions, or in accessing memory. With these abstractions, execution speeds in the thousands of device cycles per second can be achieved. Operating on top of this environment, hardware/software cosimulation packages allow the engineer to run a software debugger, the ISA software, and an RTL simulator simultaneously.

Most system-level hardware and software bugs can be detected and fixed at this stage. To complete software debug, it may be necessary to develop an even more abstract set of models to improve simulation speed further. In our example, we could substitute a C++ model for the RTL model of the data transformation block, and achieve very high simulation speeds.

To complete hardware debug, however, we need to lower our abstraction level back to RTL. The lack of detail in our ISA/behavioral models undoubtedly masks some bugs. At this point, we can run some real code on the RTL system model and perhaps some random code as well for testing unforeseen sequences. But simulation speed prohibits significant amount of real application code from being run at the RTL level.

During this debug phase, as we run application code on a high-level model and targeted tests on the RTL model, the bug rate follows a predictable curve. Typically, the bug rate increases during the first part of this testing, reaches a peak, and then starts declining. At some point on the downward slope, simulation-based testing is providing diminished returns, and an alternate method must be found to detect the remaining bugs.

11.5 Random Testing

For many chip designs, random testing has become a key verification technique, just as it has for macro-level verification. In the canonical design, Figure 11-2 on page 245, it might be useful to generate random data packets on the USB port while the processor is executing a program. The packets can be constrained to have a certain distribution in data types, target addresses, and data contents, so that they are legal values for interacting with the program running on the processor.

Another example of random testing is shown in Figure 11-4. This chip is a (very simplified) version of a network switching chip. One of the most useful methods of testing such a chip is to generate random packets at the inputs to the chip and to use the monitors to make sure the packets arrive at the appropriate output. Again, the randomization of the packets can be constrained to have an appropriate distribution of different packet types, values and destinations.

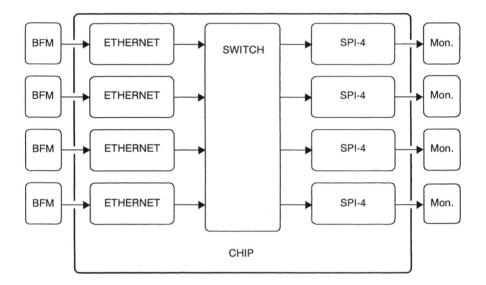

Figure 11-4   Random testing setup

This kind of random testing tends to find extreme corner case design errors much faster, and for much less effort, than directed tests. Typically the chip verification team will start random testing once finding bugs becomes more time consuming than fixing them. It is much easier for a verification engineer to debug a test failure for a test he has written than for a random test. So in the early stages of verification, directed tests are very efficient; bugs are easy to find, and easy to fix.

At some point, the design becomes robust enough that additional bugs are very time consuming to find. Writing additional tests for extreme corner cases becomes inefficient. Instead, the verification team turns to random tests, willing to incur the additional cost in debug time to shrink the effort to find bugs. This typically is at a point where the random tests run for a half hour or so without failing. The verification team may chose to continue random testing until the tests run for days or even weeks without a failure.

The biggest technical challenge in random testing is not the generation of random packets, but the efficient checking and tracing of results. One technique for checking results in an design such as Figure 11-4 on page 248 is for the BFM to send a message to the appropriate monitor when it issues a packet for that output port. Checking the design consists of having the monitor check that the packet eventually arrives at the destination port uncorrupted.

Debugging the case where a packet does not arrive can be quite challenging. To assist debug, it is useful to put monitors at intermediate points throughout the chip—for example, at the input and the output of the switch block. This approach can help identify where in the chip the packet is lost or corrupted.

11.6 Application-Based Verification

For most design teams, a key goal is to have first silicon be fully functional. This goal has motivated the functional verification plan and simulation strategies. To date, most teams have been fairly successful. According to some estimates, about 90% of ASIC designs work right the first time, although only about 50% work right the first time in the system. This higher failure rate probably results from the fact that most ASIC design teams do not do system-level simulation.

With the increasing gate count and complexity of SoC designs, it is not clear that the industry can maintain this success rate. Assume that, in a 100K gate design with today's verification technology, there is a 10% chance of a serious bug. Then for a 1M gate design, consisting of ten such modules comparably verified, the probability of no serious bugs is:

$$P_{\text{bug-free}} = .9^{10} = .35$$

Design reuse can also play an important role. If we assume that a 1M gate design consists of ten 100k blocks, with two designed from scratch (90% chance of being bug-free) and eight reused (for the purpose of discussion, 98% chance of being bug-free), then for the overall chip:

$$P_{\text{bug-free}} = .9^2 * .98^8 = .69$$

But to achieve a 90% probability of first-silicon success, we need to combine design reuse with a verification methodology that will either get individual blocks to a 99% or allow us to verify the entire chip to the 90% level.

11.6.1 Software-Driven Application Testbench

A limited amount of application-based verification can be done using simulation. As Figure 11-5 shows, the actual software application is the source of commands for the PCI bus functional model. This application can run on the workstation that is running the simulator; device driver calls that would normally go to the system bus are redirected through a translator to the simulator, using a programming language interface such as Verilog's PLI. A hardware/software cosimulation environment can provide an effective way to set up this testbench and a convenient debug environment.

The actual transactions between the application and the PCI macro under test are a small percentage of the cycles being simulated; many cycles are spent generating inputs to the bus functional model. Also, real code tends to repeat many of the same basic operations many times; extensive testing of the macro requires the execution of a considerable amount of application code. Thus, software-driven simulation is an inherently inefficient test method, but it does give the opportunity of testing the macro with real code. For large macros, this form of testing is most effective if the simulation is running on a very high-speed simulator, such as a cycle-based simulator or an emulator.

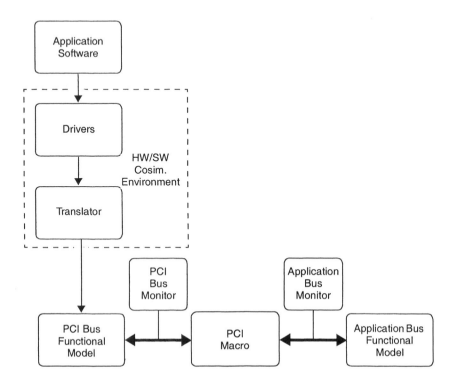

Figure 11-5   Software-driven testbench for macro-level testing

11.6.2 Rapid Prototyping for Testing

Running significant amounts of real application code is the only way to reach this level of confidence in an SoC design. For most designs, this level of testing requires running at or near real time speeds. The only available technologies for achieving this kind of performance involve some form of rapid prototyping.

The available options for rapid prototyping include:

- FPGA or LPGA prototyping
- Emulation-based testing
- Real silicon prototyping

FPGA and LPGA Prototyping

For small designs, it is practical to build an FPGA or Laser Programmable Gate Array (LPGA prototype of the chip. FPGAs have the advantage of being reprogrammable, allowing rapid turnaround of bug fixes. LPGA prototypes can achieve higher gate counts and faster clock speeds, but are expensive to turn. Multiple iterations of an LPGA design can be very costly, but can be done quickly, usually within a day or two.

Both FPGAs and LPGAs lag state-of-the-art ASIC technologies in gate count and clock speed by significant amounts. They are much more appropriate for prototyping individual blocks or macros than for prototyping SoC designs.

A number of engineering teams have used multiple FPGAs to build a prototype of a single large chip. This approach has at least one major problem: the interconnect is difficult to design and almost impossible to modify quickly when a bug fix requires repartitioning of the design between devices.

Rapid prototyping systems address this problem by using custom, programmable routing chips to connect the FPGAs. This routing can be performed under software control, providing a very flexible fast prototyping system.

Emulation Based Testing

Emulation technology grew out of attempts to provide a better alternative to using a collection of FPGAs for rapid prototyping of large chips. They provide programmable interconnect, fixed board designs, relatively large gate counts, and special memory and processor support. Recent developments in moving from FPGAs to processor-based architectures have helped to resolve partitioning and interconnect problems.

Emulation can provide excellent performance for large-chip verification if the entire design can be placed in the emulation engine itself. If any significant part of the circuit or testbench is located on the host, there is significant degradation of performance.

For our canonical design, we need to provide emulation-friendly models for the RAM, microprocessor, BFMs, monitor, and sequence generator/checker. Developing these models late in the design process can be so time consuming as to negate the benefit of emulation. It is much better to consider the requirements of emulation from the beginning of the project and to work with the memory and hard macro providers to provide these models. Similarly, the requirements of emulation must be considered in the design of the BFMs and monitors.

If executed correctly, emulation can provide simulation performance of one to two orders of magnitude less than real time, and many orders of magnitude faster than simulation.

Silicon Prototyping

If an SoC design is too large for FPGA/LPGA prototyping and emulation is not practical, then building a real silicon prototype may be the best option. Instead of extending the verification phase, it may be faster and easier to build an actual chip and debug it in the system.

To some extent this approach is just acknowledging the fact that any chip fabricated without running significant amounts of real code must be considered a prototype. That is, there is a high probability that engineering changes will be required before release to production.

The critical issue in silicon prototyping is deciding when one should build the prototype. The following is a reasonable set of criteria:

- The bug rate from simulation testing should have peaked and be on its way down.
- The time to determine that a bug exists should be much greater than the time to fix it.
- The cost of fabricating and testing the chip is on the same order as the cost of finding the next n bugs, where n is the anticipated number of critical bugs remaining.
- Enough functionality has been verified that the likely bugs in the prototype should not be severe enough to prevent extensive testing of other features. The scenario we want to avoid is building a prototype only to find a critical bug that prevents any useful debug of the prototype.

There are a number of design features that can help facilitate debug of this initial prototype:

- Good debug structures for controlling and observing the system, especially system buses
- The ability to selectively reset individual blocks in the design
- The ability to selectively disable various blocks to prevent bugs in these blocks from affecting operation of the rest of the system

11.7 Gate-Level Verification

The final gate-level netlist must be verified for both correct functionality and for timing. A variety of techniques and tools can be used for this task.

11.7.1 Sign-Off Simulation

In the past, gate-level simulation has been the final step before signing off an ASIC design. ASIC vendors have required gate-level simulation and parallel test vectors as part of signoff, using the parallel vectors as part of manufacturing test. They have done this even if a full-scan methodology was employed.

Today, for 100K gate and larger designs, signoff simulation is typically done running Verilog simulation with back-annotated delays on hardware accelerators. Running full-timing, gate-level simulations in software simulators is simply not feasible at these gate counts. Even with hardware accelerators, speeds are rarely faster than a few hundred device cycles per second.

RTL sign-off, where no gate-level simulation is performed, is becoming increasingly common. However, most ASIC vendors still require that all manufacture-test vectors submitted with a design be simulated on a sign-off quality simulator with fully back-annotated delay information and all hazard checking enabled. Furthermore, they require that these simulations be repeated under best case, nominal case, and worst case conditions. This has the potential to be a resource intensive task.

This requirement is rapidly becoming problematic for the following reasons:

- Thorough, full-timing simulation of a million-gate ASIC is not possible without very expensive hardware accelerators, and even then it is very slow.
- Parallel vectors typically have very low fault coverage (on the order of 60%) unless a large and expensive effort is made to extend them. As a result, they can be used only to verify the gross functionality of the chip.
- Parallel vectors do not exercise all the critical timing paths, for the same reason they don't achieve high fault coverage. As a result, they do not provide a sufficient verification that the chip meets timing.

As a result of these issues, the industry is moving to a different paradigm. The underlying problems traditionally addressed by gate-level simulation are:

- Verification that synthesis has generated a correct netlist, and that subsequent operations such as scan and clock insertion have not changed circuit functionality
- Verification that the chip, when fabricated, will meet timing
- A manufacturing test that verifies that the chip is free of manufacturing defects

These problems are now too large for a single solution, such as gate-level simulation. Instead, the current methodology uses separate approaches to address each issue:

- Formal verification is used to verify correspondence between the RTL and final netlist.

- Static timing analysis is used to verify timing.

- Some gate-level simulation, either unit-delay or full timing, is used to complement formal verification and static timing analysis.

- Full-scan plus BIST provides a complete manufacturing test for functionality. Special test structures, provided by the silicon vendor, are used to verify that the fabricated chip meets timing and other analog specifications.

11.7.2 Formal Verification

Formal verification encompasses both property checking and equivalence checking. At this point in the design cycle, equivalence checking is used to compare the gate-level netlist to the RTL for a design. Because it uses a static, mathematical method of comparison, formal verification requires no functional vectors. Thus, it can compare two circuits much more quickly than can be done with simulation, and with much greater accuracy.

Equivalence checking tools compare two designs by reading them into memory and then applying formal mathematical algorithms on their data structures. The designs can be successfully compared as long as they have the same synchronous functionality and correlating state holding devices (registers or latches). The two circuits are considered equivalent if the functionality is the same at all output pins and at each register and latch.

Formal verification can be used to check equivalence between the original RTL and:

- The synthesized netlist

- The netlist after test logic is inserted. For scan, this is quite straightforward; for on-chip JTAG structure, some setup is required, but the equivalence can still be formally verified.

- The netlist after clock tree insertion and layout. This requires comparing the hierarchical RTL to the flattened netlist.

- Hand edits. Occasionally engineers will make a last-minute hand edit to the netlist to modify performance, testability, or function.

One key benefit of formal verification is that it allows the RTL to remain the golden reference for the design, regardless of modifications made to the final netlist. Even if the functionality of the circuit is changed by a last minute by editing the netlist, the same modification can be retrofitted into the RTL and the equivalence of the modified RTL and netlist can be verified.

For large designs, formal verification between the gate-level design and the RTL can be too slow, especially for multiple iterations. In such cases, it is better to use formal verification once between the RTL and the gate-level netlist, then use that gate-level netlist as the golden reference for future iterations. For example, you can use formal verification to compare gate-level netlists before and after clock tree insertion. Formal verification algorithms work more efficiently when comparing gates to gates than when comparing gates to RTL.

11.7.3 Gate-Level Simulation with Full Timing

Gate-level simulation complements formal verification. Dynamic simulations are rarely an exhaustive test of equivalence, but simulation is necessary to validate that an implementation's behavior is consistent with the simulated behavior of the RTL source. Gate-level simulation is particularly important for verifying initialization because gate-level simulation handles propagation of unknown (X) or uninitialized states more accurately than RTL simulation.

On large chips, gate-level simulation with full timing is very slow, and should be used only where absolutely necessary. This technique is particularly useful for validating asynchronous logic, embedded asynchronous RAM interfaces, and single-cycle timing exceptions. In a synchronous design, these problem areas should not exist, or should be isolated so they are easily tested.

These tests should be run with the back-annotated timing information from the place and route tools, and run with hazards enabled. They should be run with worst case timing to check for long paths, and with best-case timing to check for minimum path delay problems.

11.8 Specialized Hardware for System Verification

Design teams have long recognized the limitations of software simulators running on workstations. Simulation has never provided enough verification bandwidth to do really robust system simulation. Over the last fifteen years there have been numerous efforts to address the needs of system simulation through specialized hardware systems for verification.

Early efforts focused on hardware accelerators. Zycad introduced the first widely-available commercial accelerators in the early 1980s; in the early 1990s, IKOS introduced competitive systems based on somewhat similar architectures. The Zycad systems provided very fast fault simulation; at the time fault, simulation of large chips was not really practical with software simulators. These systems were also used for gate-level system simulation. IKOS systems focus exclusively on system-level simulation.

These accelerators map the standard, event-driven software simulation algorithm onto specialized hardware. The software data structures used to represent information about gates, netlists, and delays are mapped directly into high-speed memories. The algorithm itself is executed by a dedicated processor that has the simulation algorithm hardwired into it. A typical system consists of anywhere from 4 to over 100 of these processors and their associated memory. These systems are faster than workstations because each processor can access all the needed data structures at the same time and operate on them simultaneously. Additional performance results from the parallel execution on multiple processors.

The introduction of FPGAs in the 1980s made possible another kind of verification system: emulation. These systems partition the gate-level netlist into small chunks and map them onto FPGAs; they use additional FPGAs to provide interconnect routing. These systems can execute many orders of magnitude faster than hardware accelerators. Large circuits that run tens of cycles per second on software simulators might run hundreds or a few thousand of cycles per second on a hardware accelerator. These same circuits run at hundreds of thousands of cycles per second on emulation systems.

Emulation systems achieve their high performance because they are essentially building a hardware prototype of the circuit in FPGAs.

Emulation systems, however, have a number of shortcomings:

- They operate on gate-level netlists. Synthesis is typically used to generate this netlist from the RTL. Any part of the circuit that is coded in non-synthesizable code, especially testbenches, must run on the host workstation. This considerably slows emulation. A circuit with a substantial part executed on the workstation may run as much as two orders of magnitude slower than one with the entire circuit in the emulator.

- The partitioning of the circuit among numerous FPGAs, and dealing with the associated routing problems, presents a real problem. Poor utilization and routing inefficiencies result in the need for very large numbers of FPGAs to emulate a reasonably sized chip. The resulting large systems are very expensive and have so many mechanical components (chips, boards, and cables) that they tend to experience reliability problems.

- The use of FPGAs tends to make controlling and observing individual nodes in the circuit difficult. Typically, the circuit has to be (at least partially) recompiled to allow additional nodes to be traces. This makes debugging in the emulation environment difficult.

The first problem remains an issue today, but important progress has been made on the second and third problems. New systems, have moved from a pure FPGA-based system to a custom chip/processor-based architecture. Where previous systems had arrays of FPGAs performing emulation, the new systems have arrays of special purpose processors. These processor-based systems usually use some form of time-slicing: the processor emulates some gates on one cycle, additional gates on the next cycle. Also, the interconnect between processors is time-sliced, so that a single physical wire can act as several virtual wires. This processor-based approach significantly improves the routing and density problems seen in earlier emulation systems.

The processors used on these new systems also tend to have special capabilities for storing stimulus as well as traces of nodes during emulation. This capability helps make debug in the emulation environment much easier.

These new systems hold much promise for addressing the problems of high-speed system verification. The success of these systems will depend on the capabilities of the software that goes with them: compilers, debuggers, and hardware/software cosimulation support. These systems will continue to compete against much less expensive approaches: simulation using higher levels of abstraction and rapid prototyping.

The rest of this chapter discusses emulation in more detail.

11.8.1   Accelerated Verification Overview

Figure 11-6 shows the major components of an emulation system. The components are:

Models

RTL blocks and soft IP are synthesized and mapped onto the emulation system hardware. Memory blocks are compiled and emulated on dedicated memory emulation hardware.

Physical environment

Hard macros (IP cores) that have bonded-out chips, can be mounted on special board and interfaced directly to the rest of the emulation system. Similarly, hardware testbenches, such as signal generators, can be connected directly to the emulation system.

In-circuit verification

The emulation system can be interfaced directly to a board or system to provide in-circuit emulation. Thus, an application board can be developed and debugged using an emulation model of the chip.

System environment

A software debug environment and a hardware/software co-simulation environment provide the software support necessary for running and debugging real system software on the design.

Testbenches

Behavioral RTL testbenches can be run on the host and communicate with the emulation system. Note that running any significant amount of code on the host will slow down emulation considerably.

Stimulus

Synthesizable testbenches can be mapped directly onto the emulation system and run at full emulation speeds. Test vectors can be stored on special memories in the emulation system, so that they too can be run at full speed.

These components combine to provide all the capabilities that designers need to verify large SoC designs including:

- RTL acceleration
- Software-driven verification at all levels in the design cycle
- In-circuit verification to ensure the design works in context of the system
- Intellectual property support

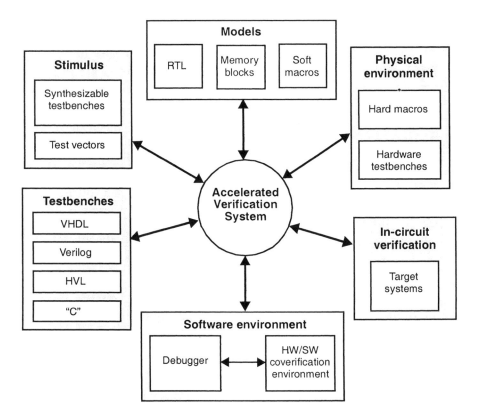

Figure 11-6 Emulation process

11.8.2 RTL Acceleration

Designers continue to use software simulators to debug their designs, but a threshold is reached where simulator performance becomes a major bottleneck for functional verification at the RTL level, especially for large SoC designs. This threshold will vary based on the design team and the verification environment. As the RTL functional simulations reach duration of more than 6-8 hours, it will become more efficient to compile the design and run it in an emulator. As an example, an RTL design that may only take several minutes to compile on a simulator, but runs for eight hours, may compile in 30 minutes on the emulator and run in a matter of seconds.

Thus, at some point in the design cycle, system simulation and debug may be more appropriately done on the emulator than on the simulator. For this debug strategy to be effective, however, we need an RTL symbolic debug environment that provides users with complete design visibility, real time variable access, and support for enumerated types and state variable assignments.

11.8.3   Software Driven Verification

As the software content in SoC designs increases and design cycles shrink, hardware/software co-development and software-driven verification become increasingly important. Software-driven verification plays two key roles in an SoC verification strategy:

- Verification of the hardware using real software
- Verification of the software using (an emulation model of) real hardware, well before the actual chip is built

Traditionally, using the software to verify the hardware has been confined to very late in the design cycle using breadboards, or early prototype runs of silicon. With reusable hardware and software blocks, it is possible to assemble an initial version of the RTL and system software very quickly. With the new emulation systems, it is possible to test this software on an emulation model of the hardware running at near-real-time speeds. Incremental improvements to both the hardware and software can then be made and tested, robustly and quickly.

In particular, the high performance of emulation systems allows the design team to:

- Develop and debug the low-level hardware device drivers on the virtual prototype with hardware execution speeds that can approach near real-time
- Boot the operating system, initialize the printer driver, or place a phone call at the RTL phase of the design cycle

11.8.4   Traditional In-Circuit Verification

As the design team reaches the end of the design cycle, the last few bugs tend to be the most challenging to find. In-circuit testing of the design can be key tool at this stage of verification because the ultimate verification testbench is the real working system. One large manufacturer of routers routinely runs its next design for weeks on its actual network, allowing the router to deal with real traffic, with all of the asynchronous events that are impossible to model accurately.

In the case of systems that operate with the real asynchronous world, with random events that might occur only on a daily or weekly basis, complete information capture is essential. The emulation system provides a built-in logic analyzer that records every signal in the design during every emulation run, using a powerful triggering mechanism. This debug environment allows designers to identify and correct problems without having to repeat multiple emulation runs.

11.8.5 Design Guidelines for Emulation

Most of the guidelines for emulation are identical to guidelines for design reuse listed in Chapter 5. These include:

Guideline – Use a simple clocking scheme, with as few clock domains as possible. Emulation works best with a fully synchronous, single-clock design.

Guideline – Use registers (flip-flops), not latches.

Guideline – Do not use combinational feedback loops, such as a set-reset latch in cross-coupled gates.

Guideline – Do not instantiate gates, pass transistors, delay lines, pulse generators, or any element that depends on absolute timing.

Guideline – Avoid multicycle paths.

Guideline – Avoid asynchronous memory.

The following guidelines are requirements specific to emulation:

Guideline – Hierarchical, modular designs are generally easier to map to the emulation hardware than flat designs. The modularity helps reduce routing between processors.

Guideline – Large register arrays should be modeled as memories, to take advantage of the special memory modeling hardware in the emulation system.

11.8.6 Testbenches for Emulation

To realize the benefits of emulation, virtually all of the circuit and testbench for the design must run on the emulator. This means that the testbench should be synthesizable.

One approach would be to make the testbench synthesizable from the beginning, and to use the same testbench for both RTL verification and for emulation. We believe that this approach is flawed.

The behavioral testbenches that can be created with current testbench automation tools are significantly more powerful than any synthesizable testbench. The capabilities for stimulus creation and automated response checking are essential for RTL test and debug, and cannot easily be replicated in the synthesizable subset of HDLs.

Instead, we recommend that a new, synthesizable, and relatively simple testbench be used for emulation. Once a bug is found, the circuit (and its current state) can be moved back to RTL simulation for debug.

If we take our canonical design, the following approach seems reasonable. In Figure 11-7 on page 263, the software for the processor is compiled and loaded into memory in the emulator. This allows the processor and peripherals to run at full emulation speed.

The stimulus for the data transformation block is also loaded into memory on the emulator. In this case, since it is an MPEG2 decoder, we can store a bit stream that represents encoded video data. A simple state machine (marked "SM") transfers data from the stimulus memory to the I/O interface. Similarly, the serial data from the output of the MPEG2 decoder is sent to a response capture memory in the emulator. Another simple state machine handles the handshake for the data transfer.

Although this approach requires a second testbench to be built, including two state machines, this is significantly less incremental effort than requiring the RTL testbench to be synthesizable.

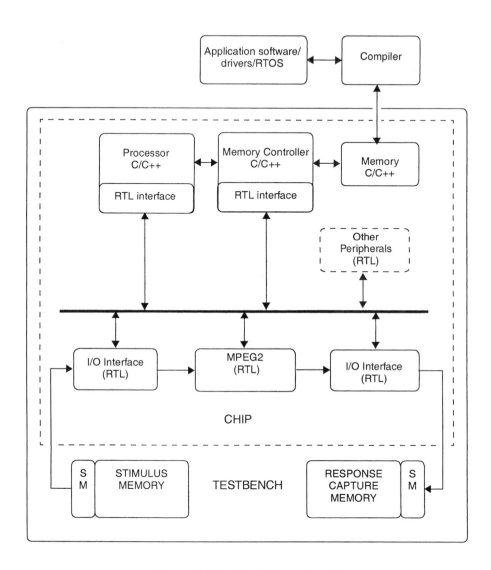

Figure 11-7 Emulation testbench

CHAPTER 12   *Data and Project Management*

This chapter discusses tools and methodologies for managing the design database for macro design and for system design. The topics are:

- Data management
- Project management

12.1 Data Management

Data management issues include revision control, bug tracking, regression testing, managing multiple sites, and archiving the design project.

12.1.1 Revision Control Systems

A strong revision control system is essential for any design project. A good revision control system allows the design team to:

- Keep all source code (including scripts and documentation) in one centralized repository that is regularly backed up and archived
- Keep all previous versions of each file
- Identify, quickly, changes between different revisions of files
- Take a snapshot of the current state of the design and label it

RCS, SCCS, and ClearCase are examples of tools with which revision control systems can be built. Additional scripts and processes are typically used to create a complete revision control system.

The most common paradigm is for each designer to be able to check out the entire design structure and recreate it locally, either by copying files or creating pointers to them. The designer then works and tests locally before checking the design files back into the central repository.

There are two different models for controlling this check-in process: the always-broken and always-working models.

The Always-Broken Model

In the *always-broken* model, each designer works and tests locally and then all the designers check in their work at the same time. The team then runs regression tests on the whole design, fixing bugs as they appear.

There are two problems with this model. First, when regressions tests fail, it is not clear whose code broke the design. If there are complex inter-dependencies between the modules, debugging regression failures can be difficult and time consuming.

The second problem with this model is that there tends to be a long integration period during which the design is essentially broken. No new design work can be done during this integration and debug phase because designers cannot check out a known-working copy of latest version of the design.

The Always-Working Model

The *always-working* model overcomes the major problems presented by the always-broken model. For the initial integration of the design, when separate modules are first tested together, the always-working model is the same as the always-broken model. Everyone checks in the initial version of the blocks and a significant debug effort ensues. In some designs, it may be possible to integrate a subset of the whole design, and then add additional blocks once the first subset is working. This approach greatly reduces the debug effort.

Once an initial baseline for the design is established, the always-working model uses the following check-in discipline:

- Only one designer can have a given block checked out for editing.
- When a block is being checked in, the entire central repository is locked, blocking other designers from checking modules in.
- The designer then runs a full regression test with the existing design plus the modified block.
- Once the regression tests pass, the designer checks in the block and removes the lock.

This model ensures that the entire design in the central repository always passes regression testing; that is, it is always working. It also greatly reduces the debug effort because only one new module at a time is tested.

We recommend the always-working model of revision control.

12.1.2 Bug Tracking

An effective bug tracking system is essential for rapid design of complex blocks and systems. A central database that collects all known bugs and desired enhancements lets the whole team know the state of the design and prevents designers from debugging known problems multiple times. It also ensures that known problems are not forgotten, and that any design that is shipped with known bugs can include documentation for the bugs.

Another key use for bug tracking is bug rate tracking. In most projects, the bug rate follows a well-defined curve, reaching a peak value early in the integration phase and then declining as testing becomes more robust. The current bug rate and the position on this curve help define the most effective testing and debug strategy for any phase of the project, and help determine when the chip is ready to tape out.

Formal bug tracking usually begins when integration begins; that is, as soon as the work of two or more designers is combined into a larger block. For a single engineer working on a single design, informal bug tracking is usually more effective. However, some form of bug tracking is required at all stages of design.

12.1.3 Regression Testing

Automated regression testing provides a mechanism for quickly verifying whether changes to the design have broken a previously-working feature. A good regression testing system automates the addition of new tests, report generation, and distribution of simulation over multiple workstations. It should also highlight differences in output files between passing and failing tests, to help direct debug efforts.

12.1.4 Managing Multiple Sites

Many large projects involve design teams located at multiple sites, sometimes scattered across the globe. Effective data management across these sites can facilitate cooperation between the teams.

Multiple site data management starts with a high-speed link between sites; this is essential for sharing data. The revision control central repository must be available to all sites, as well as bug tracking information. Regression test reports must be available to all sites.

The key to managing a multi-site project is effective communication between the individual engineers working on the project. Email, voicemail, and telephones have been the traditional tools for this. Technology is now readily available for desktop video conferencing with shared displays and virtual whiteboards. All of these techniques are needed to provide close links between team members.

One management technique that helps make these technology-based solutions more effective is to get the entire team in one place for an initial planning and teambuilding session. Once team members get to know the people at other sites, the daily electronic communication becomes much more effective.

12.1.5   Archiving

At the end of any design project, it is necessary to archive the entire design database, so that it can be re-created in the future, either for bug fixes or for enhancements to the product. All aspects of the design must be archived in one central place: documentation, source code, all scripts, testbenches, and test suites. All the tools used in the design must also be archived in the revision control system used for the design. If these tools are used for multiple projects, obviously one copy is enough, and the tool archives can be kept separate from the design archive.

The observation above may seem obvious, but let me interject a personal note. Several years ago, I was hired to do the next generation design of an existing large system. The first thing I did was to try to find the people who had worked on the previous generation design and to find the design archive.

Well, the designers had all moved on to other companies. The system architect was still with the company but busy with another project, and he wasn't all that familiar with the detailed design anyway. The design files were scattered across a wide variety of machines, some of which were obsolete machines cobbled together from spare parts and whose disks were not backed up! Worse than that, I found several copies of the design tree, with inconsistent data. It was impossible to collect even a majority of design files and know with confidence that they were the latest versions.

In the middle of the previous design effort, the team had changed silicon vendors and tools. The HDL for the design was *almost* accurate, but some (unspecified) changes were made directly to the netlist. This scenario is the manager's nightmare. It took months to recover the design archive to the point where the new design effort could begin; in fact, close to half the design effort consisted of recreating the knowledge and data that should have been archived at the end of the design.

12.2 Project Management

There are many excellent books on project management [1,2,3], and we will not attempt to cover the subject in any depth. However, there are a several issues that are worth addressing.

12.2.1 Development Process

An ISO 9000-like development process, where processes are documented and repeatable, can help considerably in producing consistently high-quality, reusable designs. Such a process should specify:

- The *product development lifecycle*, outlining the specific phases of the design process and the criteria for transitioning from one phase to another
- What *design reviews* are required at the different stages of the design, and how the design reviews will be conducted
- What the *sign-off process* is to complete the design
- What *metrics* are to be tracked to determine the completeness and robustness

Two key documents are used to communicate with the rest of the community during the course of macro design. These documents are the *project plan* and the *functional specification*. These are both living documents that undergo constant modification during the project.

12.2.2 Functional Specification

A key characteristic of a reusable design is a pedigree of documentation that enables subsequent users to effectively use it. The requirements for a functional specification are outlined in Chapter 4. This specification forms the basis for this pedigree of documentation, and includes:

- Block diagrams
- Functional specification
- Description of parameters and their use
- Interface signal descriptions
- Timing diagrams and requirements
- Verification strategy
- Synthesis constraints

In addition to the basic functional information above, it is quite useful to keep the functional specification as a living document, which is updated by each user throughout its life. For each use of the block, the following information would be invaluable to subsequent generations of users:

- Project it was used on
- Personnel on the project
- Verification reports (what was tested)
- Technology used
- Actual timing and area results
- Revision history for any modifications

12.2.3 Project Plan

The project plan describes the project from a management perspective and documents the goals, schedule, cost, and core team for the project. Table 12-1 describes the contents of a typical project plan.

Table 12-1   Contents of a project plan

Part	Function
Goals	Describes the business reasons for developing the macro and its key features and benefits, including the technical goals that will determine the success or failure of the project.
Schedule	Describes the development timeline, including external dependencies and risks. The schedule should contain sufficient contingency time to recover from unforeseen delays, and this contingency should be listed explicitly.
Cost	Describes the financial resources required to complete the project: headcount, tools, NREs, prototype build costs.
Core Team	Describes the human resources required to complete the project: who will be on the team, who will be the team leader.

References

1. Floyd, Thomas et al. *Winning the New Product Development Battle*. IEEE, 1994.

2. McConnell, Steve. *Software Project Survival Guide*. Microsoft Press, 1997.

3. Demarco, Tom and Lister, Timothy. *Peopleware: Productive Projects and Teams*. Dorset House, 1999.

CHAPTER 13

Implementing Reuse-Based SoC Designs

In May, 1997, Gary Smith, principal EDA analyst at Dataquest, Inc. at that time, alerted the design community in his article published in Virtual Chip Design, entitled "The Revolution Isn't Coming—It's Already Here." In this article, Gary stated:

> The modern-day realization of systems-on-chips has started a revolution, one that is every bit as important to electronics as was the microprocessor revolution of the 1970s.

SoC design presents fundamental challenges in all domains: silicon technology, EDA tools, data management, IP design, and risk management.

The key enabler for SoC design, however, has been the design community's adoption of a reuse-based methodology. In previous editions of this book, we have concluded with a discussion of how to implement such a reuse-based design methodology. But this methodology is now in such wide use that we decided instead to share the experiences of some teams who have been through the implementation process.

During the preparation of the third edition, we spent many hours discussing reuse and SoC designs with many teams from many different companies throughout the world. This chapter contains a few brief selections from these conversations.

13.1 Alcatel

Thierry Pfirsch is Core Program Manager in the Hardware Coordination Division of Alcatel. His team is responsible for reuse activities in Alcatel. Thierry remarked:

> Alcatel's hardware coordination division has invested a lot of time and effort to make IP reuse a reality for our design teams. The *RMM* has been instrumental in the implementation of the process and Chapter 5 is online on our internal web site. The first step in implementing reuse was IP awareness, which has been accomplished, through a sophisticated internal web site: KMS (Knowledge Management System). This Internal Web Site allows each designer and design manager to exchange information on the actual IP used inside Alcatel and at the same time to exchange internal IP. The KMS provides an information exchange between the designers, using discussion forums, news, and so on.

> Now that IP awareness is established in Alcatel, we are working with Design & Reuse to encapsulate external IPs in order to allow the catalog and transfer functions. The goal of the encapsulation platform is to give a full visibility on what is available on the external market, and to reduce the time to buy an external IP. The internal Alcatel IP Catalog is automatically updated by the IP providers through the IP encapsulation platform from Design & Reuse and includes at the same time some confidential information for Alcatel such as price, name of the last user, and integration report.

> But the most important role of my team is to guarantee a level of quality to the end user. We have created an IP evaluation service that provides an evaluation report for each IP. Therefore the designer has from the start a reference document that helps him in the decision process and gives him an overview of the quality of the IP. The result of this type of evaluation process is not a guarantee of the functionality of the blocks, but more of risk evaluation on the use of the block.

> Alcatel is a strong participant in the VC Quality Development Working Group (VSIA) that was created late 2001. Our reuse efforts are in line with the efforts from VSIA and the OpenMORE initiative that were merged at DAC 2001.

> From an SoC design point of view, Alcatel has used a platform-based methodology for SoC for several years. We have an internal product called CleanDmep. For a system house, being able to create virtual system prototypes at the earliest stage possible of the SoC software/hardware design process gives a tremendous advantage from a time-to-market point of view. The platform is based on the ARM/AMBA subsystem, but there is no limitation on the use of the methodology of this platform.

Thierry also drew a parallel between the Intel PC design methodology and Alcatel 's SoC design methodology (Figure 13-1 on page 273):

> In both cases, you start from the star IP (processor + bus), derive the hardware platform with the peripherals and add the software contents to create your virtual prototype.

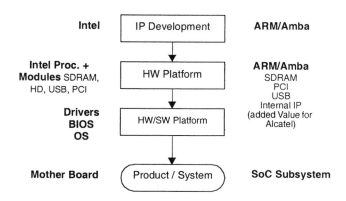

Figure 13-1 Alcatel's SoC design flow

13.2 Atmel

Erich Palm is Director, Libraries and Design Tools at Atmel, Rousset, France. He remarked:

> The overall SoC design methodology at Atmel is based on concurrent engineering, close to the Spiral SoC design flow described in the *RMM* (page 14). Figure 13-2 on page 275 shows Atmel's Concurrent SoC Design Flow, centered around the chip design. Around this, to the left is the path for the application design, and to the right, the concurrent design at the library/cell/macro level, be it digital macros, cores, custom IPs, memories, or analog. Concurrent engineering requires good coordination and formal quality checkpoints which are ensured through overall project management by the chip design team. On the other hand, well established, formalized delivery procedures are necessary from the IP/cell provider's end.

> The Atmel development model makes a clear separation between the application (software) and hardware content of the SoC. The Atmel chip design team deals with the hardware content. The external customer or the internal application group deals with the software content. FPGA prototyping is often done for digital IP blocks and ASIC prototyping is used for the chip. Since the chip verification is essentially done with hardware, there is less need for emulation based verification. On the subject of memory validation between the behavioral model and the physical implementation, simplified electrical simulations are still used for comparison with the behavioral model outputs, but formal verification techniques are becoming more necessary with increasing memory sizes. Overall, a new design based on Atmel's sophisticated IPs or subsystems takes around six to nine months, and a derivative three to four months.

Erich remarked that reuse is an internal culture at Atmel:

> IP reuse can start at any level, as indicated in the flow diagram, and is meaningful far beyond the reuse of digital IP. For example, the schematic level is often the starting point for analog cell reuse. The trend to push further design and IP reuse will continue in the future, for the sake of continuous improvement of design productivity and reliability.

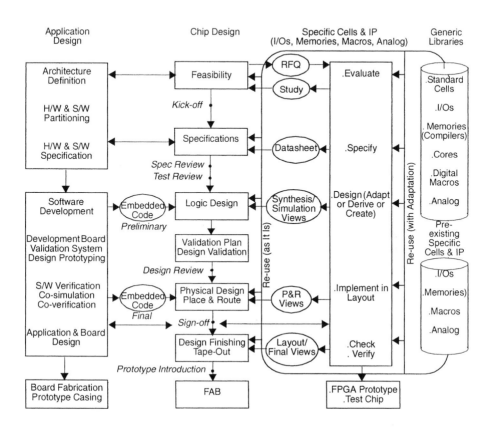

Figure 13-2   Atmel's SoC concurrent engineering SoC flow

13.3 Infineon Technologies

We talked with Albert Stritter, Vice-President, Design Automation and Technology, and Yves Saboret, Director in Albert's group working in Sophia Antipolis, France. Albert's group services the overall Infineon business lines: Wireline-Wireless-Security & Chip Card IC-Automotive & Industrial-Memory Products. Albert remarked:

> The SoC design flow is a very stable and robust flow and consists of designing completely different applications using existing IP blocks like TriCore, ARM, MIPS, Carmel, and Oak, but is strongly focused on one common methodology called InWay. InWay serves both digital and analog/mixed signal designer needs.

Yves stated:

> The digital SoC design flow has not changed much over the past few years. Integration of new tools and qualification of libraries or relative design representations for the different steps is still very engineering intensive. My major concern is the lack of top-level methodology and suitable tools to keep a consistent representation of the reusable blocks through the flow from RTL to GDSII.

Albert outlined the changes that are significant in the digital and mixed signal SoC flows:

> The quality assurance has been automated and enhanced. The major issues are coming from the new technology descriptions that make it more difficult to integrate them in the flow. Each BU has their own IP repository system and uses them intensively. The design flow from RTL to GDSII is there, and continuously improved. We currently see, however, with the advent of very deep submicron technologies, that a new paradigm shift for the complete flow is necessary. The EDA offerings are lagging behind some years, probably because these new technologies are not yet in their focus. On the other hand, these new technologies are showing effects that are not yet completely known.

Albert noted that a lot of progress has been made in the area of technology libraries:

> Qualification of libraries has been enhanced to the point where standard cells and I/Os can be generated in a few weeks from scratch. ESD structure design is pretty much automated, as well as memory design as well. In fact, we are able to design a complete library at an internal cost that is almost half of the one we can find on the market and in a time that beats everybody. Where we can still improve is the area of model generation and verification. This is still cumbersome, but we are working on it.

Albert also noted that, today, one of the major weaknesses was hierarchical verification. He commented on the fact that for IP blocks, standards for software and hardware design representations do not coexist. As for the SoC verification reuse platform, Albert added that this form of verification is application-specific, and therefore will be very difficult to standardize.

Infineon Technologies' SoC design flow is shown in Figure 13-3 on page 277.

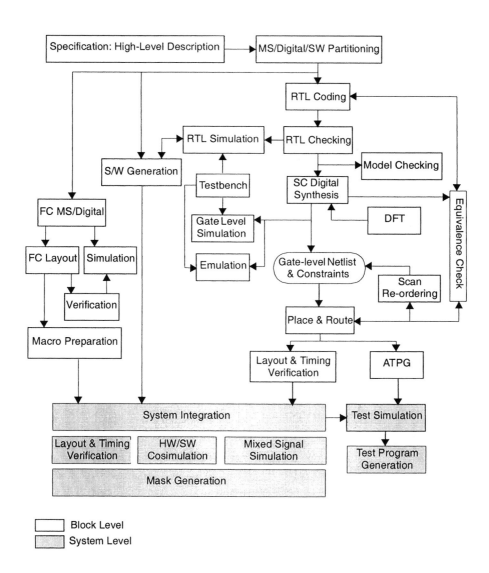

Figure 13-3 Infineon Technologies' SoC design flow

13.4 LSI Logic

We talked to Tim Daniels, Technical Product Marketing Manager for the ASIC Division of LSI Logic based in the UK. Their typical SoC design flow is shown in Figure 13-4 on page 279. The flow can be segregated into two areas:

- Front end: Design and physical planning
- Back end: Physical design and verification

A key characteristic of the flow today is the division between front end and back end which is now post timing closure/placement/global routing. Today's process technologies require that the front end flow understand and incorporate physical issues to avoid undesirable surprises in the back end flow.

Designers tend to jump into floorplanning, synthesis and placement early, either with traditional flows (synthesis and layout) or with physical synthesis tools, as a way of driving timing closure. However an effective implementation methodology starts with RTL that is sympathetic to layout and this is an area in which many flows could benefit. The majority of the industry does not incorporate RTL analysis into their SoC flows today.

RTL generation in advanced SoC flows starts with RTL coding guidelines and includes tools to analyze RTL for rule compatibility. RTL coding quality, from a timing closure concern, has the single greatest impact on physical design. A source of RTL coding rules is the *RMM*—many of which are incorporated into LSI Logic's checking rules. Examples of high-quality RTL coding include:

- Registering all module outputs
- Using 1 clock per compilation module
- Using only 1 clock edge
- Avoiding large structures and high-fanout nets
- Using consistent names
- Not having fall-through paths within a block
- Avoiding tristates, multiply driven nets and feedback loops
- Using an effective hierarchical structure
- Ensuring registers use reset

In today's SoC designs, the RTL of both the SoC design and embedded IP blocks should be thoroughly evaluated to avoid problems with physical implementation. RTL code should be analyzed as it is written, not when problems are encountered in physical synthesis or layout. Tools exist both in the RTL and the netlist space to analyze an IP block, a submodule or a whole design and are an invaluable asset in LSI Logic's Advanced SoC Design Flow.

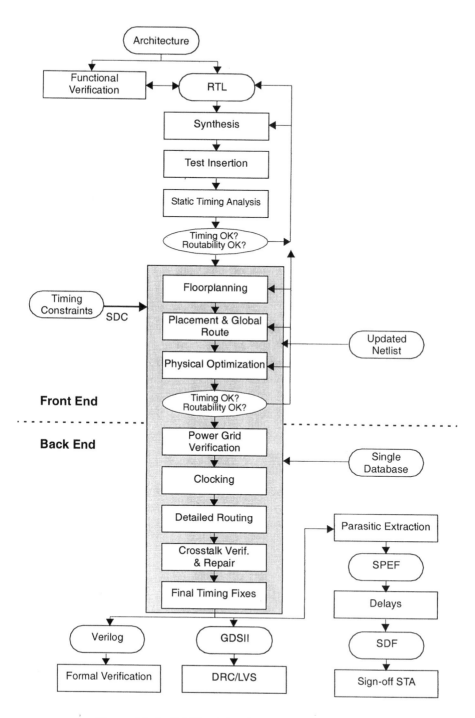

Figure 13-4 LSI Logic's advanced SoC design flow

13.5 Philips Semiconductor

We talked with Louis Quere, IC Development Manager, BL Digital Media, Pierre-Guy Margueritte, IC development Group Leader, BL Digital Media, Pierre Lieutaud, Emulation Technology Manager, Patrick Rousseau, IC Development Group Leader, and Alain Rambert IC Design Support and CAD Support Manager in the Philips Semiconductor plant in Caen, France. We discussed the SoC design flow around two types of ICs: Set Top Box based on a MIPS CPU and MPEG with Video-Audio Encoder/Decoder and derivatives. The latter IC is around 35 million transistors and an area of 104 square mm in a 0.18 micron CMOS process.

Louis stated:

> The SoC design flow is equivalent to the Spiral SoC design flow described in the *RMM* (page 14) with a very heavy emphasis on emulation for verification. An important characteristic of the group of chips was the fact that they shared a common architecture. This was considered paramount in the verification reuse strategy implementation. Another very strong remark was made that software content is more and more important and that bandwidth with the external memory is a bottleneck. Although the verification strategy is heavily based on emulation, a need for higher level of abstraction is seen especially for software modeling and the memory interface IPs.

The Business Line Digital Media team agrees that complex SoCs will rely heavily on predefined IP blocks. Philips Semiconductor 's answer to the complexity challenge is an evolution of design styles that goes from the full-custom approach to an architecture reuse approach based on the Nexperia™ Silicon System Platform (Figure 13-5 on page 281), all based on IP hardware and software reuse.

Another interesting concept that has been implemented is that any hardware module that has not been tested to specifications can be switched off during simulation. This allows advanced testing during final qualification of the IP hardware module. Another important rule is that each chip has one unified testbench at the top level. The team makes heavy use of "internal spies" for RTL top-level verification, especially for data flow monitoring.

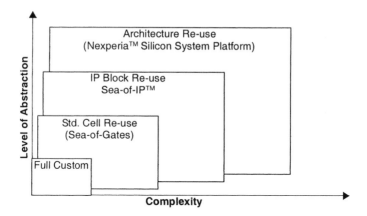

Figure 13-5   Philips Semiconductor's Nexperia Silicon System Platform

13.6 STMicroelectronics

We talked with Thierry Bauchon, Director R&D, Set Top Box Division and François Remond, Design Support Manager. They offered these comments:

> Due to the type of circuits and a design cycle involving the end customer, the overall SoC Design Flow needs to bring predictability in the design process. We do this through a top-down, block based design approach. Very 'clean' interface points are needed at each step of the design flow to avoid costly iterations. The heart of the block-based SoC design approach is a block-based floorplanning tool. The top-down approach with a block-based design style allows the classical 'divide-and-conquer' strategy, similar to what is described in the *RMM* (page 240). A typical set top box SoC consists of 15 to 30 blocks with an equal split between soft and hard macros (IP blocks).

François stressed:

> The block-based approach enables concurrent development of the top-level and the block-level design until the final integration. This design approach fits well with increasing need of emulation and validation of the overall design, which is eased by the divide-and-conquer approach.

Thierry added:

> For the designs done in cooperation with the end customer, the specific ASIC methodology, where the customer does the front-end design up to the netlist, has evolved to have the customer work at floorplan level up to the netlist generation.
>
> Another crucial choice is the system bus for the chip. IPs come from various providers: ST, customers, or third-party vendors. Compatibility between those IPs and the system bus is achieved by designing specific wrappers, or even introducing a customer bus with a bridge to the system bus. The overall SoC can have some IP blocks connected to the ST internal bus and others connected to the customer bus. This requires extremely well documented interfaces between IPs and the buses and well documented bridges between buses.

STMicroelectronics' design planning steps are shown in Figure 13-6 on page 283.

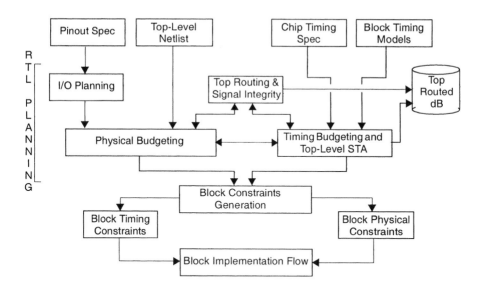

Figure 13-6 STMicroelectronics' design planning steps

13.7 Conclusion

We would like to thank all of our colleagues from around the industry who took the time to talk to us about reuse and its role in SoC design. It is clear that the SoC design challenge is well understood and that electronics companies have dedicated significant efforts and talent to meet this challenge.

It is also clear that there is a great deal of commonality in how design teams are approaching SoC design. All the teams we talked to based their SoC methodology on the reuse of key IP, especially processors, buses, interfaces, and memory.

Another common message from our discussions is that software is becoming a dominant component in the SoC. Design teams are paying close attention to new methodologies and tools for hardware-software co-design.

On the other hand, a number of people we talked to expressed concern that mixed signal integration, especially the RF in the telecommunication industry, has not exhibited the progress and innovation seen in other areas, and that this is becoming a significant challenge.

Verification continues to be a major area of interest, and concern, for SoC design. Design teams are paying close attention to new verification tools and methodology, particularly the use of verification IP such as bus functional models and monitors. On the other hand, a number of people we talked with expressed concern about the slow progress in the area of formal verification techniques such as property checking and assertion checking. Achieving 100% confidence in designs is still an unsolved problem, and these new techniques are regarded as key steps towards achieving this goal.

In summary, the design community has made remarkable progress over the last few years in SoC design and in the reuse technology that makes it possible. Methodologies and tools for reuse-based design are maturing rapidly. Unfortunately, the inexorable progress of Moore's Law means that new challenges are arising almost as fast as the old ones are solved. But that's what makes all our jobs so exciting.

Bibliography

Books on software reuse:

1. *Measuring Software Reuse*, Jeffrey S. Poulin, Addison-Wesley, 1997.
2. *Practical Software Reuse*, Donald J. Reifer, Wiley, 1997.

Formal specification and verification:

1. http://www.ececs.uc.edu/~pbaraona/vspec/, the VSPEC homepage.
2. *Formal Specification and Verification of Digital Systems*, George Milne, McGraw-Hill, 1994.
3. *Formal Hardware Verification*, Thomas Kropf (ed.), Springer, 1997.

Management processes:

1. http://www.sun.com/sparc/articles/EETIMES.html, a description of the UltraSPARC project, mentioning construct by correction.
2. *Winning the New Product Development Battle*, Floyd, Levy, Wolfman, IEEE.

Books and articles on manufacturing test:

1. "Testability on Tap," Colin Maunder et al., IEEE Spectrum, February 1992, pp. 34–37.
2. "Aiding Testability also aids Testing," Richard Quinell, EDN, August 12, 1990, pp. 67–74.
3. "ASIC Testing Upgraded," Marc Levitt, IEEE Spectrum, May 1992, pp. 26–29.
4. Synopsys *Test Compiler User's Guide*, v3.3a, 1995.
5. Synopsys *Test Compiler Reference Manual*, v3.2, 1994.

6. Synopsys *Certified Test Vector Formats Reference Manual.*

7. *Digital Systems Testing and Testable Design*, M. Abromovici et al., Computer Science Press, 1990.

8. *The Boundary Scan Handbook*, Kenneth Parker, Kluwer Academic Publishers, 1992.

9. *The Theory and Practice of Boundary Scan*, R. G. "Ben" Bennetts, IEEE Computer Society Press.

10. *Testability Concepts for Digital ICs*, Franz Beenker et al., Philips Corp, 1994.

11. "A Comparison of Defect Models for Fault Location with IDDQ Measurements," Robert Aitken, *IEEE Design & Test*, June 1995, pp. 778–787.

Books and articles on synthesis:

1. "Flattening and Structuring: A Look at Optimization Strategies," Synopsys Application Note Version 3.4a, April 1996, pp. 2-1 to 2-16.

2. *VHDL Compiler Reference Manual*, Synopsys Documentation Version 3.4a, April 1996, Appendix C.

3. *DesignTime Reference Manual*, Synopsys Documentation Version 3.4a, April 1996.

4. "Commands, Attributes, and Variables," Synopsys Documentation Version 3.4a, April 1996.

Index